Sir J. William Dawson

On Specimens of Eozoon Canadense and their Geological and Other Relations

Sir J. William Dawson

On Specimens of Eozoon Canadense and their Geological and Other Relations

ISBN/EAN: 9783337186241

Printed in Europe, USA, Canada, Australia, Japan

Cover: Foto ©berggeist007 / pixelio.de

More available books at **www.hansebooks.com**

ON SPECIMENS OF EOZOON CANADENSE AND THEIR GEOLOGICAL AND OTHER RELATIONS.

BY SIR J. WILLIAM DAWSON, LL.D., F.R.S.

I. INTRODUCTORY.

Whatever may be the ultimate decision of Palæontologists as to the nature of Eozoon, it is important that the original specimens on which its description was based and those later acquisitions which have thrown farther light on its structure and have been published in that connection, should be preserved and catalogued.

The collections made by Sir W. E. Logan are now for the most part in the Museum of the Geological Survey at Ottawa. Those accumulated by the author of these notes, as well as duplicates presented by Sir W. E. Logan, are in the Peter Redpath Museum.* It is to these latter collections that the present paper relates, and the object is to render them as useful as possible for scientific purposes in the future.

In order, however, to secure this result, it will be necessary that these notes shall not consist of a mere list of specimens; and for this reason they will include such notices of the

* Very large and valuable collections, more especially of microscopic preparations, were also in the possession of the late lamented Dr. Carpenter, of London, who was engaged in their study up to the time of his death.

geological history of Eozoon, its modes of occurrence and
states of preservation, as may enable any reader to compre-
hend the nature of the questions raised by the specimens.
Reference will be made not to the whole literature of Eozoon
which is somewhat voluminous, but to the more important
publications relating to it. Extracts will also be given
from some of the principal descriptive papers relatiug to
Eozoon and to fossils illustrating its mode of occurrence and
probable affinities, and copies will be inserted of some of the
illustrations which have been published of the specimens in
the collections.

II. GEOLOGICAL RELATIONS OF EOZOON.

These have been very fully discussed by Sir W. E. Logan,
by Dr. T. Sterry Hunt and by the writer, * and their
conclusions have been confirmed by more recent observa.
tions.

The oldest known stratified rock in Canada is the massive
orthoclase gneiss, designated by Sir William Logan the
gneiss of Trembling Mountain, and of unknown thickness.
It has since been called by Hunt the Ottawa gneiss, and is
probably equivalent to the Lewisian, Fundamental or Bo-
gian gneiss of European Geologists. In so far as
known, it is destitute of limestones and of organic remains. The great spread of this rock would seem to
indicate that it is of vast thickness. The thickness observed by Logan at Trembling Mountain was estimated at
5000 feet. This lowest portion of the Laurentian may be
named the Ottawa or Lewisian gneiss. As compared with
the succeeding formations, it is remarkable as containing no
quartzites, slates, limestones or dolomites, such as would
indicate ordinary submarine waste of rock, or evidencing
any distinction of land and water. It is therefore strictly a
fundamental rock, and may be a portion of the original crust

* Logan, Journal of Geological Society, Feb., 1865. Hunt, Geo. Canada, 1866,
p. 231 ; Amer. Journal Sci. [2] XXXVII.,431 ; XL., 369 ; Quar. Jour. Geol. Soc.,
XXI., 67. Also 2nd Geol. Survey of Penn. Rep. E., p. 168. Dawson, Life's Dawn
on Earth (London, 1875.)

of the earth formed before the ordinary causes of sedimentation prevalent in later times were inaugurated.

On this rests a second division which Hunt has termed the Grenville series. This has a thickness of about 20,000 feet, and includes, besides great masses of gneiss, beds of Limestone, of Diorite, of Pyroxene Rock, of Quartzite and of Magnetite. The Limestones, which are continuous beds traceable for long distances, occur in three principal bands, all of them of great thickness and consisting of crystalline limestone with dolomite and with disseminated Serpentine, Graphite, Apatite and Mica. They also include subordinate layers of gneiss and quartzite, and irregular beds and concretions of Pyroxene Rock. It is in the uppermost of the three limestone bands, known as the Grenville band, that the most perfect specimens of Eozoon have been found.

Above the Grenville series is found the so-called Norian or Labradorian series, the Upper Laurentian of Logan, which both in the Province of Quebec and in New York is locally unconformable to the lower series, and is remarkable for its thick beds of Gneissoid Anorthosite and Labradorite Rock, though it also contains orthoclase gneiss and beds of limestone and iron ore (Ilmenite). This series has afforded no fossils, and there is reason to believe that a large portion of its material is of igneous origin, indicating great earth-movements in the later part of the Laurentian age, with access to sources of basic material below the crust of the earth.* It indicates, however, a considerable lapse of time, and is perhaps represented locally by schistose rocks, such as some of those which in the West have been reckoned as Lower Huronian, or as the upper part of the Laurentian in Southern New Brunswick, and the Micaceous series of the Archaean in New Hampshire and Virginia.

It may be useful to reproduce here the Section given by Sir William Logan of these Laurentian Rocks on the north side of the Ottawa river, where they have been traced over a great area, and as the writer can testify from personal ob-

* Selwyn, Reports Geol. Survey of Canada, 1879-80, pages 4 and 188. Also, Preliminary Report of Mr. F. D. Adams, B·A.Sc.

servation, have been very accurately mapped by Logan and the officers of the Geological Survey.

Logan's Section of Laurentian Rocks.

(Order ascending.)

		Feet
First	Orthoclase Gneiss of Trembling Mountain (Lower or Ottawa Gneiss)	5,000 or more.
First	Limestone or Limestone of Trembling Lake	1,500
Second	Orthoclase Gneiss, between Trembling Lake and Great Beaver Lake	4,000
Second	Limestone, or Limestone of Great Beaver Lake and Green Lake, with two interstratified bands of garnetiferous rock and hornblendic orthoclase gneiss, making up about half its volume	2,500
Third	Orthoclase Gneiss, with bands of garnetiferous gneiss and Quartzite, between Beaver Lake and the Rouge River	3,500
Third	Limestone or Limestone of Grenville, in some places including a band of Gneiss: (*Eozoon Canadense*). Its thickness varies from 1,500 to 60 feet, average thickness estimated at	750
Fourth	Orthoclase Gneiss including a thin bed of Limestone (Proctor's Lake) and 600 feet of Quartzite	5,000
	Norian or Labradorian or Upper Laurentian series—estimated at	10,000
	Total	32,250

It is to be observed here that, while the Lower Gneiss formation is, so far as known, wholly of that rock, in the second or Grenville member we find also Schists, Limestones, Dolomites, Quartzites, Iron ores and Graphitic Gneisses, indicating ordinary aqueous detritus, and some distinction of land and water, and such agency of life as we find in later geological formations. To whatever kind of origin therefore we attribute the orthoclase gneisses, we trace indications of a primitive period in which rock of this kind alone was formed, whether as a portion of the original crust of the earth unmodified by any sub-aerial or marine or aqueous agency, or as a product of a special aqueo-igneous action, and this succeded by a period in which ordinary aqueous agencies and atmospheric agencies played a large part and which has continued up to the present day. I would not propose on this account to disintegrate the Laurentian system, because the presence of its peculiar gneiss characterizes the whole of it, and would rather consider the Ottawa and Grenville series as two great divisions of the Laurentian.

In the Ottawa district the tortuous outcrop of the Grenville limestone was traced by Logan and his assistants for about 200 miles, and this has since been greatly extended by his successors. In the Burgess district on the opposite side of the Ottawa, and distant about 80 miles from the limit of Logan's work, Vennor has found and mapped a similar series, with three bands of limestone separated by gneisses apparently of still greater thickness than those of the Grenville district, and as in the latter, the third limestone holds Eozoon. There is therefore no room left for doubt as to the regularly bedded character of the Laurentian rocks and the continuity of their outcrops over great distances.

The Limestones of the Laurentian are thus of great volume and of vast geographical extent, and they range in thickness from 60 to 1,500 feet, while their actual horizontal area must be enormously greater than the distance they have been traced would indicate. These limestones are also associated with gneissose and schistose beds, exactly in the same way in which palæozoic limestones are associated with sandstones and shales; and some of them are ordinary limestones, while others are more or less dolomitic, in which also they resemble the palæozoic limestones. Every geologist knows that the beds which in the succeeding geological periods are the representatives of these Laurentian limestones, are not only fossiliferous, but largely composed of the debris of oceanic organisms, and that it is to the purer and more crystalline beds that this statement most fully applies. May we not reasonably infer that the great Laurentian limestones are of similar origin?

This may be farther illustrated by the structure observed in localities explored by the writer.

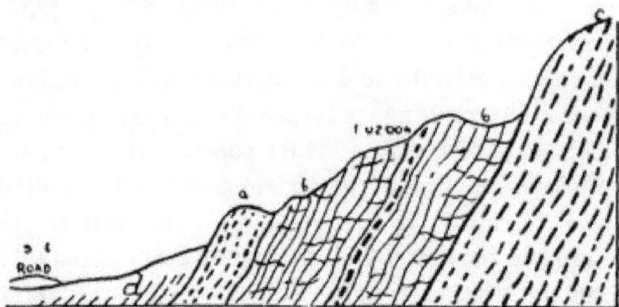

Fig. 1. Section at Cote St. Pierre.
(a) Band of Gneiss.   (b) Limestone and Dolemite, with Serpentine and Eozoon.
(c) Diorite and Gneiss.   (d) Limestone.

At Cote St. Pierre, north of Papineauville on the Ottawa River, one of the best localities of Eozoon, the limestone of the Grenville band with a high dip to the south-eastward, rests on thick-bedded diorite and gneiss of Logan's Second band, which form high ground to the north-westward. The lower part of the limestone is coarsely crystalline, with concretionary masses of white Pyroxene. Above this it becomes serpentinous and contains layers of dolomite, and in this part of the band Eozoon occurs. Above this is a thin stratum of gray gneiss, included in the limestone, which beyond this point is for the most part concealed by alluvial soil, and by a small lake, but where it appears is seen to contain Eozoon in some of its layers. Along the same line of outcrop to the south-west, the limestone holds in its lower part and associated with Eozoon, small crystalline aggregations of bluish apatite. In the best section at this place, about 150 feet of the thickness of the limestone is exposed, but this is evidently only a small portion of its entire amount. The best specimens of Eozoon occur in layers only a few feet in thickness, and perhaps in two portions of the bed; but fragmental Eozoon is found in other layers.

The following notes from a paper published in the Quarterly Journal of the Geological Society of London (1876) give some details as to the Cote St. Pierre locality:

Cote St. Pierre, in the Seigniory of Petite Nation, on the river Ottawa, is the locality whence some of the most instructive specimens of *Eozoon* were obtained by the late Mr. Lowe, whose collections are referred to in the original papers by Sir W. E. Logan, Dr. Hunt and the writer. Believing that a re-examination of this place would afford a good opportunity for collecting additional specimens, and for the study of the fossil *in situ*, as well as for testing the validity of objections recently raised to the animal nature of *Eozoon*, I made arrangements for visiting it in September, 1875 ; and, through the kindness of Mr. Selwyn, Mr. T. C. Weston, of the Geological Survey, a skilful collector, and who has had much experience in preparing and examining specimens of *Eozoon*, was permitted to accompany me, and subsequently prepared slices and photographs of some of the specimens obtained.

The Lower Laurentian rocks of this region have been carefully mapped and described in the Reports of the Geological Survey, to which I may refer for their general description. The limestone, which has afforded *Eozoon* at Cote St. Pierre, is a thick bed belonging to the Grenville band of Sir W. E. Logan, and included between the two great belts of orthoclase gneiss (the third and fourth gneiss) which in this region constitute the upper beds of the Lower Laurentian. Its average thickness, according to the measurements of Sir William Logan, is 750 feet ; but it varies from 1,500 feet to 60 feet. Its outrop has been traced in the country north of the Ottawa for at least 200 miles, along several anticlinal and synclinal folds*.

At Cote St. Pierre this limestone occurs on the flank of a hill of gneiss and stratified diorite, with a dip to the south-east at angles of 70° to 80°. The dip, however, is very inconstant, owing to the contortions of the beds.

The limestone is white and crystalline, and may be described as thin-bedded, since it presents a great number of layers of no great individual thickness, and differing in the coarseness of the crystalization and in the presence of dolomite, serpentine, and layers of gneissose matter in some of them. The specimens of *Eozoon* were found to be abundant in only one bed, not more than four feet in thickness, though occasional specimens and layers of fragments occur in other parts of the band. The exposures are in part natural weathered surfaces seen on a wooded bank, in part an opening

---

* See map in Geology of Canada, 1863.

made by Mr. Lowe to extract specimens of *Eozoon*, and a larger opening made, as we were informed, by parties in search of fibrous serpentine, or "rock cotton," for economic purposes.

The sections seen in the artificial exposures may be tabulated as follows, though from the highly inclined position of the beds and the irregularity of the excavations, perfect accuracy was not attainable :—

Mr. Lowe's excavation (order descending).

1. Limestone with serpentine and entire specimens of *Eozoon* 3 feet.
2. Coarse crystalline limestone, with layers containing fragments of *Eozoon*—4 feet.
3. Limestone with concretions and layers of serpentine, and a few specimens of *Eozoon*—several feet, to the bottom of the excavation.

Fisher's excavation (order descending).

1. Laminated limestone with bands of serpentine—6 feet.
2. White laminated limestone traversed by small veins of chrysotile—8 feet.
3. Limestone with concretions and interrupted bands of serpentine and pyroxene, and fragments of *Eozoon*—10 feet. (Crystals and layers of dolomite occur in this and the preceding beds.)
4. Limestone with large concretions of serpentine, and in one layer fine-grained variety of *Eozoon* (var. *minor*)—20 feet.
5. Limestone with serpentine and perfect specimens of *Eozoon* (This probably corresponds to Lowe's excavation)—12 feet.
6. Coarse-grained limestone and dolomite—several feet.

(After a break of several yards)

7. Limestone with masses of pyroxene and veins of chrysotile and some imperfect *Eozoon*.

(After a break of several yards)

8. Coarse-grained diorite, resting on a thick band of gneiss.

In front of Lowe's excavation, and apparently overlying the limestone exposed in it, is a narrow ledge of fine-grained gneiss; and beyond this, other and probably overlying limestone appears, holding pyroxene and mica. The whole vertical thickness of the limestone exposed can scarcely exceed 150 feet; but this is probably only a small part of the development of the band at this place.

In the strike of the limestone to the N.E. it appears to bend abruptly, or to be thrown by a fault, to the south-east, the gneiss and diorite coming forward into line with it, and the limestone appearing at a less angle in a little bare knoll in front of these. On the surface of this limestone were found some fine specimens of weathered *Eozoon*. A short distance farther to the northeastward there is another section opened by quarrying operations, in which similar beds holding *Eozoon* are exposed, and in one of the larger limestones bluish apatite occurs in disseminated grains.

I examined carefully the relation of the bedded serpentine and the veins of chrysotile or fibrous serpentine to the limestone. The compact serpentine is evidently an original part of the deposit occurring in layers and lenticular concretions. In some beds it shows no indication of the structure of *Eozoon*; but in others it fills the cavities of the fossils, and there are many regular layers of fragmental *Eozoon* of considerable thickness in which it fills the cells, while in other layers interstratified with these, the fossil is associated with dolomite. I satisfied myself on this point not only on the ground, but also by taking away large specimens representing several thin layers, and treating them with dilute acid, so as to bring out the structure. The following is a section of such a specimen,* 5½ inches in vertical thickness, treated with acid and examined with a lens :—

1. Limestone with crystals of dolomite and a few fragments of *Eozoon*.
2. Fine-grained limestone with granules of serpentine—the latter filling the chamberlets of fragments of *Eozoon* and small globigerine Foraminifera, or fragments of acervuline *Eozoon*.
3. Limestone with dolomite and including a thin layer of serpentine as above.
4. Limestone and dolomite with grains of serpentine and fragments of supplemental skeleton of *Eozoon*.
5. Crystallized dolomite, holding a few fragments of *Eozoon* in the state of calcite.
6. Limestone with disseminated serpentine as above, chamberlets of *Eozoon* and fragments of its supplemental skeleton, also small groups of acervuline chamberlets.

In other specimens a like thickness of rock presented a mass of fragments of supplemental skeleton with the canals injected with serpentine, and granules of the same filling the chambers.

The chrysotile veins, which are sometimes an inch or more in thickness, but branch off into the most minute films, are evidently altogether subsequent in origin to the bedded limestone and serpentine. They are undoubtedly of aqueous origin, and in their mode of occurrence strongly resemble the veins of fibrous gypsum penetrating the Lower Carboniferous marls of Nova Scotia. They cross

* This specimen is now in the Museum.

the bedding in all directions, and pass through the structure of *Eozoon*, though sometimes running parallel to its laminæ for short distances. They must have been introduced after the *Eozoon* was mineralized, and have evidently no connection with its structure.

I have no hesitation in stating that the assertion that these chrysotile veins are identical with or similar to the proper wall of *Eozoon* either in structure or distribution, is wholly without foundation, other than that which may arise from confounding dissimilar structures accidentally associated with each other.

Some slickensided joints lined with a lamellar and fibrous serpentine traverse the beds, and, as the chrysotile veins sometimes terminate in them, may be older than the latter. These also were observed to cross the masses of *Eozoon*.

Few disseminated minerals, other than those already mentioned, were observed in the *Eozoon* limestone. A few detached crystals of mica, pyroxene, and pyrite were found in the fragmental layers, and also a few rounded particles of quartz, probably grains of sand.

The perfect examples of *Eozoon*, at least those rendered evident by mineralization with serpentine, are confined to certain bands of limestone, and notably to one band—that originally opened by Mr. Lowe. In this bed the fossil occurs in patches of various sizes, some of them two feet or more in diameter, and bent or folded by the contortions of the strata; others are much smaller, down to a few inches. On the weathered surfaces the specimens mineralized with serpentine project, and exhibit their lamination in great perfection, resembling very closely the silicified *Stromatoporæ* of the Niagara and Corniferous Limestones.

None of the specimens of *Eozoon* are of any great vertical thickness. The lower laminæ are generally the best developed and with the thickest supplemental or intermediate skeleton. The upper laminæ become thin-walled, though often very regular; and after a great number of these laminæ the superficial portion becomes acervuline or vesicular and then terminates. In some exceptional specimens only a few laminæ have been formed. In others they become very numerous. A very fine and regular specimen (in the Museum collection) has about one hundred laminæ in a thickness of $3\frac{1}{2}$ inches, giving a little more than a thirtieth of an inch for each lamina of sarcode and test.

Many of the best specimens in the Museum are from this locality. Some of them were presented by the late Sir W. E. Logan, others were collected by myself in several visits, and others are from collections made for the Museum by Mr. E. H. Hamilton, B.A.Sc.

At another locality on the Calumet River in the Township of Grenville, the limestone of the Grenville Band is traceable in a very regular manner for several miles on the two sides of a minor anticlinal included in a larger synclinal form. It is here probably at least 600 feet in thickness, and forms a depression running between the second and third gneiss bands, which rise in ridges on either side. In this section, which I have studied in the 4th concession of Grenville, its lower part is seen to consist of a highly crystalline limestone, destitute so far as known of *Eozoon*. Above this is a very regular band of gneiss dividing the limestone into two portions, and in the limestone a short distance above this, *Eozoon* occurs in considerable quantity, but not so well preserved as at Côte St. Pierre.

Whatever views may be entertained as to the genesis of the beds of gneiss and other siliceous rocks alternating with the limestones, there can be no doubt that the latter are regular aqueous beds, as regular in their original deposition as any limestones of later date, though they have been locally much contorted and faulted by movements subsequent to their deposition. Farther, the serpentine occurring in these limestones as well as the disseminated graphite and apatite must have been deposited contemporaneously with the limestone. It is also to be observed that the serpentine of these Laurentian limestones is a peculiar variety of the mineral of a lighter colour and of lower specific gravity than ordinary serpentines, and containing "less oxyde of iron and more water."[*] It is also to be noted that the occurrence of dolomite in the limestones shows the presence of magnesia in the waters in which they were deposited. The limestones are often traversed by true veins of chrysotile or fibrous serpentine and by other veins carrying apatite, graphite with pyroxene, mica and other minerals. These vein-formed deposits are of later date than the limestones which they traverse.

The writer has formed very definite opinions as to the manner and causes of the deposition of the Laurentian

* Hunt, Geology of Canada, p. 471.

gneisses and their contained beds, and as to the reasons of their
difference in composition and texture from more modern
deposits; but may content himself with stating here that
the whole of the evidence points to the conclusion that they
are marine beds, formed however in connection with conditions of the earth's crust and of the ocean, somewhat different from those prevailing in the subsequent geological
periods, after those great movements of the crust which
ridged and folded and altered the Laurentian sediments before the beginning of the Palæozoic Period.

I have referred in the above pages to the thickness of the
Laurentian as very great. This is, however, to be taken rela-
tively to the succeeding formations and not to the earth itself.
In reality the whole ascertained thickness of the Laurentian
cannot be estimated at more than five or six miles. It comes
therefore within those narrow limits assigned by Davison,
Darwin and Mellard Reade to the superfical crust of compression and folding in a contracting globe. That the Laurentian beds have been thus compressed and folded even before
the deposition of the next succeeding formations, no one who
has studied them can doubt. On the other hand there are
physical facts which show that they are very thin in comparison with the under-crust on which they have, as it
were, slid in the process of compression. One of these is
the existence of long fissures filled with igneous rock running
across the crumpled Laurentian for great distances in perfectly direct lines. For example, on Sir W. Logan's map
of the Ottawa district, there is indicated one of these dykes,
extending westward from St. Jerome, crossing indifferently
all members of the Laurentian in every attitude, and which
the writer has seen preserving its position and direction as
far west as Templeton, more than an hundred miles from its
first appearance at St. Jerome. Such a dyke shows that
the folded Laurentian rocks are of inconsiderable thickness,
compared with the uniform and continuous under-crust on
which they rest, and which cracks altogether without reference to them, just as a very thin veneer of corrugated wood
laid on a thick plank would be obliged to crack in conform-

ity with the fissures in the latter. The Laurentian beds, with all their crumplings and mountainous elevations, may thus be compared to a mere wrinkled scum on the surface of a pot of melted metal, or to a thin veneer on the surface of the under-crust which has no such structure. The importance of this consideration, with reference both to the original stratification of the Laurentian and to theories of mountain-making, is obvious.

III. STATE OF PRESERVATION.

We may first ask, under this head, what are the structures supposed to be preserved. On the supposition that Eozoon was a marine organism, its test or hard part, which grew on the sea bottom, consisted of a series of calcareous laminæ, not perfectly parallel, but bending towards each other at intervals, and uniting so as to form flattened chambers, deeper toward the base and becoming shallower in the upper part, while at the top they sometimes become broken up into rounded cells or chamberlets, constituting an "acervuline" mass. The chambers, which, on the supposition above stated, were originally filled with the sarcodic matter of the animal, were after death and the burial of the skeleton in some calcareous sediment, occupied with mineral substances introduced by infiltration, and more especially with serpentine and pyroxene, which were at the same time being deposited in layers and concretions in the surrounding material. When well preserved, the calcareous laminæ are seen to be traversed with innumerable canals, terminating in very fine tubuli. These canals are occupied by serpentine, pyroxene or dolomite, or by limestone, according to the state of preservation. (See Figs. 2, 3, 4).

The masses of Eozoon sometimes consist of as many as one hundred and fifty laminæ superimposed. Originally flat or rounded, they assumed in growth club-shaped or turbinate forms, and sometimes by coalescence formed wide sheets or irregular masses, in which case they are often observed to be traversed in their thickness by conical or cylindrical

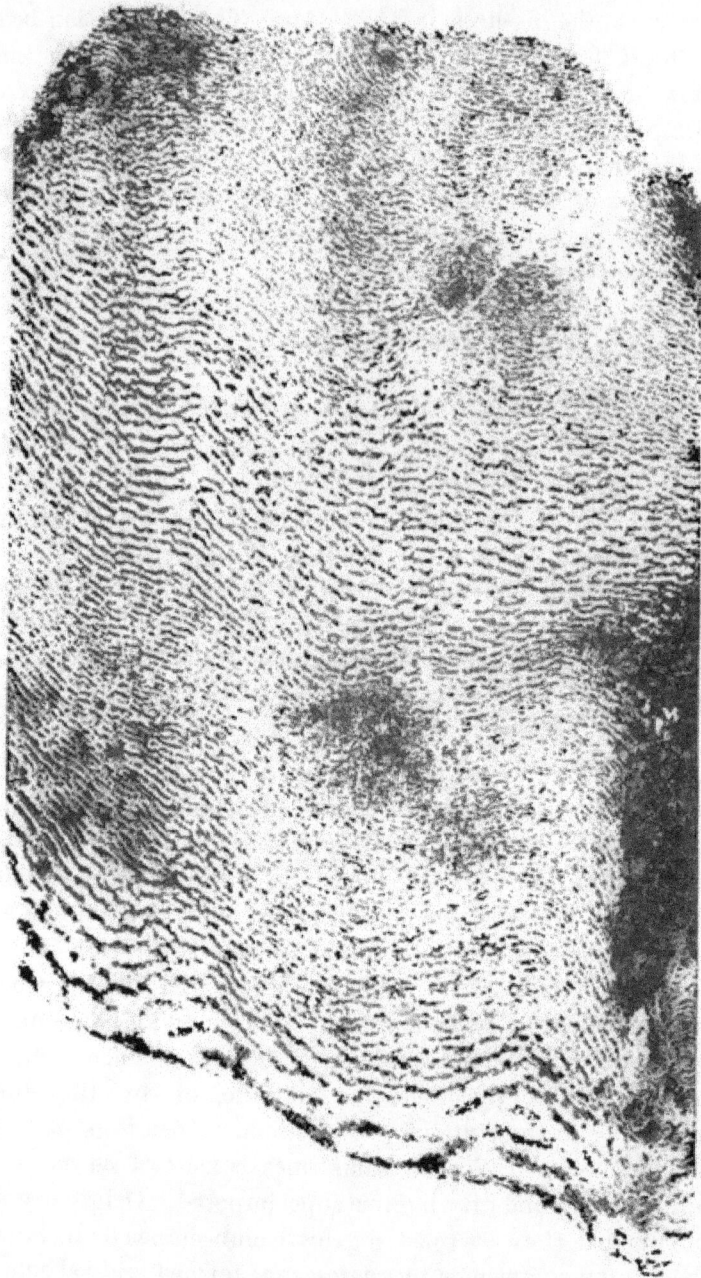

Fig. 2. Nature-printed specimen of Eozoon.

tnbes or oscula. (See frontispiece.) The outer surface and the walls of these tubes were strengthened by bending and coalescence of the laminæ. The mode of growth would be similar to that of more modern organisms of the genera *Loftusia, Carpenteria* and *Polytrema*, and to that of some kinds of *Stromatoporæ*. Finally, these calcaceous tests were liable to be broken up and scattered in fragments over the sea bottom, constituting the material of beds of organic limestone, like the coral sand that surrounds modern reefs and islands.

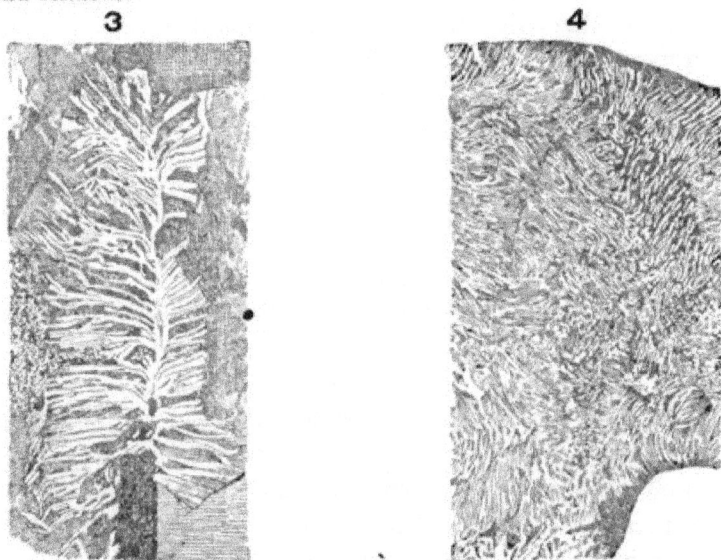

FIG. 3. Canal System of Eozoon injected with serpentine (magnified).
FIG. 4. Very fine canals and tubuli filled with Dolomite (magnified).
(From Micro-photographs.)

Assuming Eozoon to be a fossil animal of the characters above described, its mode of preservation in the ordinary serpentinous specimens is more simple than that of many fossils of later date. The calcareous walls have remained substantially unchanged, except that they have become somewhat crystalline in structure, and in many cases have assumed the crystalline cleavage of calcite; but this change is quite common in Palæozoic shells and crinoids. The chambers have been filled and the canals and tubuli traversing the calcareous test have been injected with a hydrous silicate. This is a filling up by no means infrequent in later fossils, and as Dr. Carpenter has shewn, it is going on in the modern seas in the case of foraminifera and other porous tests and shells injected with glauconite. Numerous instances of this kind exist in Palæozoic limestones. Several of these are described in my paper on fossils mineralized by silicates (*Jour. Geol. Society,* Feb. 1879, *et infra*), and I have recently met with another interesting example in a limestone from the Lower Carboniferous of Maxville, Ohio, collected by Prof. E. B. Andrews. and presented to me by Dr. T. Sterry Hunt, in which many crinoids and corals are beautifully injected with a greenish hydrous silicate resembling glauconite.

Mineralization of this kind is in reality greatly less complex than that in which, as in many fossil corals and fossil woods, the calcarerous or woody matter has been entirely removed and replaced by silica, oxyde of iron or pyrite. In many cases also in Palæozoic fossils the cavities have been filled with successive coats of different minerals giving very complex appearances. I have in my collection a specimen of Stigmaria in which every vessel has been coated in the interior with successive linings of red and white calcite, and subsequently filled with calcite and pyrite, and in a Sternbergia from the coal formation the phragmata are silicified and encrusted with crystalline silica and pyrite, while the interstices are filled in with sulphate of barium. Such complex and eccentric examples of fossilization are much more intricate than anything that occurs in the ordinary examples of Eozoon.

Geologists should also be reminded that porous fossils, once infiltrated with siliceous minerals, are practically indestructible. Nothing short of absolute fusion can wholly deface their structures, and these remain in many cases in the utmost perfection when the external forms have been wholly lost or inseparably united with the matrix.

There is therefore nothing anomalous in the preservation of Eozoon, except its occurrence in rocks highly crystalline and of unusually great age; and but for these circumstances it is probable that no doubt would have been entertained on the subject. The question of the crystalline structure of rocks containing fossils deserves, however, some further consideration.

That in limestones a crystalline condition is compatible with the preservation of fossils, and more especially with the preservation of their microscopic characters, is very well known. Many Palæozoic limestones are of a highly crystalline character, and yet retain abundant evidence of their organic origin. For example, the Chazy and Trenton limestones of the vicinity of Montreal have a perfectly crystalline fracture, and present to the naked eye no trace of any form but cleavage planes of calcite, yet, when sliced and studied with the microscope, they are seen to consist of organic fragments having their most minute structures preserved, but so completely enveloped and identified with the crystalline calcite which fills their pores and interstices that they cleave with it. It is to be observed also that in these limestones, instances occur in which organic fragments are inscribed in hexagonal crystals and might be mistaken for mere crystals containing impurities, did not these latter show on examination the original structures. Mesozoic and even Tertiary limestones have sometimes assumed the same conditions. That the Laurentian limestones holding Eozoon have undergone no change incompatible with the preservation of fossils, is proved by the fact that they still retain their original lamination, and present layers, often quite thin, of dolomite and calcite, and of the latter with various mixtures of serpentine, graphite, &c. Now there is no reason why the structures of any fossil should not survive when the lamination of the limestone remains.

Another example quite in point is that of some large calcified trees of the coal period. When broken, these trunks show large coarse cleavable crystals like those of stalagmite, but when sliced it is often found that the structure has been perfectly preserved in the midst of the crystallization.

That the laminæ of Eozoon themselves are in some

2

cases replaced by dolomite, or partially by flocculent serpentine, is no argument against their organic nature. Stromatoporæ, shells and corals are often found to have their calcareous material wholly or in part replaced by other minerals, as dolomite, carbonate of iron, pyrite and silica. The replacement by the latter mineral more especially gives us many of our most beautifully preserved Palæozoic fossils. At Pauquette's Rapid on the Ottawa, among the numerous fossils found in a silicified state imbedded in the limestone, are many Stromatoporæ, and in these the layers are not merely filled but actually replaced with silica, which, while it retains the form of the laminæ is itself arranged in curious concretionary grains which might at first sight be mistaken for a part of the structure.

In the Silurian dolomite of Guelph in Ontario, specimens of Cœnostroma, replaced by perfectly crystalline dolomite, not only show their lamination, but in some cases even their fine canals. In the gray dolomite of Niagara, similar appearances are observed. In some places it is filled with masses of Stromatopora dispersed through the dolomite just as Eozoon is in the Laurentian limestone. These fossils are silicified and vary in diameter from a foot to an inch. The greater part are spheroidal in form, but some are cylindrical or club-shaped, while others spread into flat sheets or are of various irregular shapes. In many specimens, the structure is beautifully preserved; but in others it has partially disappeared, and the substance of the fossil is replaced by coarsely crystalline calcite or dolomite, or presents cavities lined with crystals of these minerals. There is reason to believe that many cavities in the limestone, now empty and coated with these crystals, were once occupied by Stromatoporæ, or by the species of sponge found in this limestone. In every respect, except in the absence of hydrous silicates, the mode of occurrence of these fossils resembles that of Eozoon at Côte St. Pierre.

In some such cases of replacement it is probable that the original material of the fossil was arragonite, and for this reason more easily removed or replaced. Every Palæontologist is familiar with the fact that arragonite or prismatic shell has been removed in cases where lamellar shell has remained, and the latter has sometimes disappeared when compact calcite shells, like those of Balanus, for example, have escaped. In the case of Eozoon, however, as in that of foraminifera in general, the calcite seems to have been of the less perishable kind, and this may be connected with the integrity of the calcareous wall in the better preserved specimens.*

By what appears to a palæontologist a strange perversion of reasoning, some of the opponents of the organic nature of Eozoon take the badly preserved specimens as typical, and suppose that these represent an original mineral condition, which in the better preserved specimens has only assumed its greatest perfection.

As I have often urged, this kind of argument would invalidate all reasoning from the structures of fossils. In all large masses of fossil coral or wood, we find portions in all stages of disintegration. Sometimes the centre is a mere structureless mass, when the surface is perfectly preserved; Sometimes it is the surface that is disorganised. In other cases portions are well preserved, and others disintegrated in the most capricious manner. I have specimens of fossil coniferous wood in which portions are disintegrated along the medullary rays, giving the appearance of widely separated wedges, and others in which concentric bands are alternately preserved and destroyed, others in which irregular spaces have been eaten out and filled with structureless matter, and others in which crystalline or concretionary structures have been developed in spots, giving the most grotesque and inexplicable appearances. Yet in all these cases we have the general form of a trunk and portions of it in which the structures are preserved. In one example of silicified wood I have found regularly formed prisms of quartz deposited in rows along the woody fibres as if these had formed original parts of the structure.

---

* I have elsewhere remarked that the calcareous wall of Eozoon retains a *finely granular* texture, similar to that seen in shells, etc., in altered Palæozoic limestones.

In fossil woods it is also very common to find the tissues compressed, folded and contorted in spots, so as to give the most unnatural possible appearances. Now in all such cases it is surely reasonable to take the well–preserved portion as the means of interpreting the rest, though I have known cases where, for want of attention to this, portions of woody tissue have been described as cellular, in consequence of their being disintegrated by the crystallization of quartz.

It is also to be observed that there is a gradation in the probability of the preservation of structures. A very finely tubulated structure, like that which is supposed to have constituted the proper wall of Eozoon, is rarely perfectly preserved. In modern foraminifera infiltrated with glauconite, we usually see their finer structures preserved only in spots, or a part of the length of the tubes only filled. The larger cells are often infiltrated when the tubuli are empty. A coarse canal system is more likely to be perfectly infiltrated. Further, in Tertiary Nummulites the fine tubes are often filled with calcite, while the glauconite has penetrated the coarser portions only. This is very well seen in the beautiful specimens from Kempfen in Bavaria. All this applies to Eozoon. The most difficult part to find is its proper wall. The coarser canals are often present without the finer. The coarser parts of the canals are sometimes filled with serpentine, when the finer branches are filled with calcite or dolomite. The cells and laminæ are sometimes quite manifest when the finer structures are absent. All this is in perfect harmony with the analogy of other fossils.

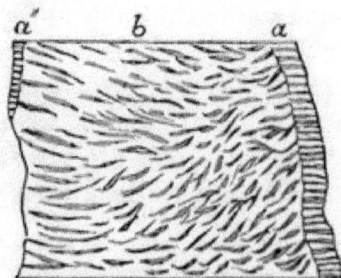

Fig. 5.   Slice of single lamina of Eozoon, magnified.   (*a*) Tubulated wall ; (*b*) Canal system ; both injected with Serpentine.

Eozoon also agrees with other fossils in the independence of its form with reference to the mineral matter with which the cavities may be filled. This peculiarity commended itself to the sagacity of Sir William Logan, and induced him to argue for the organic nature of Eozoon before its minute structures were known, and since these were investigated the argument has been much strengthened. The minerals serpentine, pyroxene and loganite are found filling the chambers, and the two former with dolomite and calcite occupy the canals, which often present calcareous fillings in the finer ramifications, when the main stems are occupied with serpentine. These facts are readily explained if we assume cavities and tubes of definite form to be filled with minerals according to circumstances; but they are not explicable on the supposition of a merely inorganic origin. They correspond perfectly with facts observed in the infiltration and replacement of all classes of fossils, which often occur in such a way that similar spaces are occupied in one part of the fossil with one mineral, in others with another.

In connection with this, the imperfections in the preservation of Eozoon are also parallel with those observed

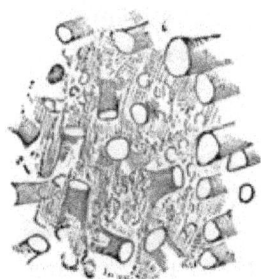

Fig. 6. Cross section of canals, injected with serpentine, highly magnified.

in different organic substances. As an example, I have already mentioned that in some of the specimens a white flocculent serpentine encroaches upon the calcareous walls or in part replaces them. This would indicate the partial removal of the calcite prior to or at the same time with the filling. In some cases also the calcite wall is wholly or in part replaced with dolomite. Such changes are not infrequent in Palæozoic fossils in which the substance of a calcareous part has often been wholly removed and replaced by another mineral or has been partially eroded and so in part replaced.

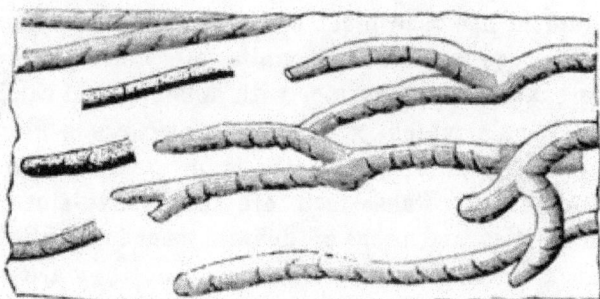

Fig. 7.   Longitudinal section of canals, highly magnified.

There are other peculiarities deserving special notice:—

1. In some specimens the serpentine filling the chambers presents a laminated appearance, as if deposited in successive layers. There even occur serpentine-lined cavities and canals with calcareous filling. This may depend on the deposition of serpentine in coatings on the sides of those cavities, leaving perhaps a central portion to be filled with calcite, or may in some cases be the result of the filling of the cavities with successive laminæ of serpentine from below upward. In either case we have frequent examples of these varieties of filling in ordinary fossils.

2. There are examples of Eozoon in which no serpentine or other mineral filling appears, and in which the whole mass is calcareous, though presenting canals filled with serpentine or dolomite. In these cases the explanation is that the mass of Eozoon has not had its cavities filled, but has been compressed by pressure into a solid mass. Such a state of preservation is often observed in other fossils, more especially in fossil wood, in which the cell-walls often become under pressure wholly coalescent.

3. The condition of the proper wall also illustrates the manner of preservation. The tubes which compose it are so

extremely fine that they are rarely injected with silicates. Sometimes they are merely occupied with calcite, and in this case the wall constitutes an apparently structureless band, or merely presents a band of slightly different appearance from the remainder. Sometimes the tubuli appear as fine continuations of the canals ; or as a more or less perfect fringe of fine lines, and in decalcified specimens, this part is often represented merely by a tabular space between the ends of the canals and the serpentine filling. In the best specimens and in very thin slices under a high power, these tubuli appear as hollow threads with expanded terminations, but this is rarely to be seen. All these conditions may be equally well observed in Nummulites injected with glauconite.

4. The larger masses of Eozoon have often suffered considerable contortion and even faulting, and this seems to have occurred in some instances previous to complete fossilization. This is a condition often observed in fossils of all ages, and every palæontologist is familiar with the fact that in all the older formations even the hardest calcareous fossils have been affected with accidents of this nature.

There are even a few examples in the collections which would seem to indicate that portions had been broken off, perhaps by the action of the waves, previous to fossilization. It is not unlikely that some of the specimens have been loose and subject to the action of the waves and currents before being imbedded.

5. An interesting feature in connection with the specimens of Eozoon from St. Pierre, noticed in previous papers, is the occurrence of layers filled with little globose casts of chamberlets, single or attached in groups, and often exactly resembling the casts of Globigerinæ in greensand. On weathered surfaces they were often especially striking when examined with the lens. In some cases, the chamberlets seem to have been merely lined with serpentine, so that they weather into hollow shells. The walls of these chamberlets have had the same tubulated structure as Eozoon ; so that they are in their essential characters minute acervuline specimens of that species, and similar to those I described in my paper of 1867 as occurring in the limestones of

+ 50

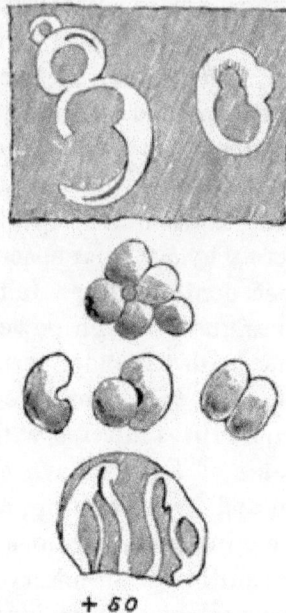

Fig. 8. Sections and casts of detached chomberlets, magnified.

Long Lake and Wentworth, and also in the Loganite filling
the chambers of specimens of Eozoon from Burgess. Some
of them are connected with each other by necks or processes,
in the manner of the groups of chamberlets described by
Gümbel as occurring in a limestone from Finland, examined
by him. That they are organic I cannot doubt, and also
that they have been distributed by currents over the surface
of the layers along with fragments of Eozoon. Whether

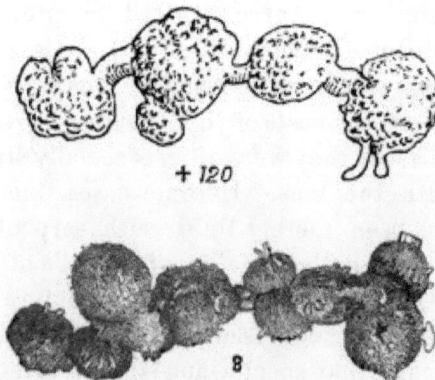

+ 120

8

Fig. 9. Groups of chamberlets, Canada and Finland, magnified.

they are connected with that fossil or are specifically distinct, may admit of more doubt. They may be merely minute portions detached from the acervuline surface of Eozoon, and possibly of the nature of reproductive buds. On the other hand they may be distinct organisms growing in the manner of Globigerina. As this is at present uncertain, and as it is convenient to have some name for them, I have proposed to term them Archæosphærinæ, understanding by that name minute Foraminiferal organisms, having the form and mode of aggregation of Globigerina, but with the proper wall of Eozoon.

A specimen in the collections from Cote St. Pierre deserves notice (Fig. 11 *infra*) as illustrating the nature of Archæosphærinæ. It is a small or young specimen, of a flattened oval form, $2\frac{1}{2}$ inches in its greatest diameter and of no great thickness. It is a perfect cast in serpentine, and completely weathered out of the matrix, except a small portion of the upper surface, which was covered with limestone which I have carefully removed with a dilute acid. The serpentinous casts of the chambers are in the lower part regularly laminated; but they are remarkable for their finely mammilated appearance, arising from their division into innumerable connected chamberlets resembling those of Archæosphærinæ. In the upper part the structure becomes acervuline, and the chamberlets rise into irregular prominences, which in the recent state must have been extremely friable, and, if broken up and scattered over the surfaces of the beds, would not be distinguishable from the ordinary Archæosphærinæ. This specimen thus gives further probability to the view that the Archæosphærinæ may be for the most part detached chamberlets of Eozoon, perhaps dispersed in a living state and capable of acting as germs. Other specimens weathered out and showing granular forms have been collected by Mr. E. H. Hamilton and are now in the Museum.

6. Specimens of Eozoon have been traversed by veins of chrysotile and calcite which cross all their structures indifferently, and often seriously affect their preservation. But

similar accidents have affected fossils of every age, and especially those of the older and more altered rocks. The manner in which these veins cross the forms of Eozoon in truth present an additional proof that these are original enclosures in the limestone, and not products of any subsequent change.

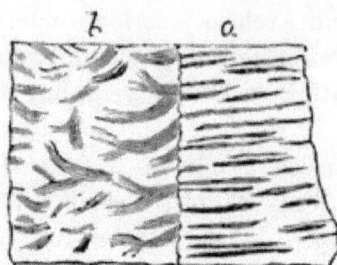

Fig. 10. Chrysotile vein crossing Eozoon, magnified. (a) Vein of fibrous Serpentine or Chrysotile ; (b) Tubulation of Eozoon.

7. In connection with this I would refer to a fact which I have often previously mentioned, namely, that the Laurentian limestones, when destitute of the laminated forms characteristic of Eozoon, are nevertheless often filled with small patches showing the minute structures. These I regard as fragments of Eozoon broken up and scattered by the currents. In this case, the remainder of these bands of limestone must be composed of fragments of other organisms which not being porous have not been so preserved by infiltration as to be distinguishable. In the original investigation of Eozoon, however, a great number of slices of these fragmental limestones were prepared by Mr. Weston the lapidary of the Geological Survey, and carefully examined, and though they showed no distinct structure except that of Eozoon, I felt convinced, and expressed this conviction in my original description, that these fragments presented such traces of structure as one is familiar with in metamorphosed organic limestones of more modern date.* At Côte St. Pierre there are several layers of limestone and dolomite studded

* Especially the finely granular structure above referred to.

with this fragmental Eozoon, and in specimens from Brazil, from Warren County, New York, and from Chelmsford in Massachusetts, and St. John, New Brunswick, the traces of Eozoon which I have observed consist of these fragments.

8. In slicing one of my specimens from Côte St. Pierre, I have recently observed a very interesting peculiarity of structure, which deserves mention. It is an abnormal thickening of the calcareous wall in patches extending across the thickness of four or five lamellæ, the latter becoming slightly bent in approaching the thickened portion. This thickened portion is traversed by regularly placed parallel canals of large size, filled with dolomite, while the intervening calcite presents a very fine dendritic tubulation. The longitudinal axes of the canals lie nearly in the plane of the adjacent laminæ. This structure reminds an observer of the *Canostroma* type of *Stromatopora*, and may be either an abnormal growth of Eozoon, consequent on some injury, or a parasitic mass of some stromatoporoid organism finally overgrown by the Eozoon. The structure of the dolomite shows that it first incrusted the interior of the canals, and subsequently filled them—an appearance which I have also observed in some of the larger canals filled with serpentine, and which is very instructive as to their true nature.

The above statements have reference to state of preservation, and are intended to remove misconceptions on that subject, but the mere fact of so many coincidences both in state of preservation and defects and imperfections between Eozoon and ordinary fossils, furnishes in itself, independently of other evidence, no small proof of its organic origin.

IV. NEW FACTS AND SPECIAL POINTS.

Under this heading, I shall summarise some of the previous statements, and add some special facts bearing on the character of the specimens and their interpretation.*

* Nos. 1 to 11 were read at the Meeting of the British Association, Sept. 5, 1387, and printed in part in *Geological Magazine*, February, 1888.

(1.) Form of Eozoön Canadense.

Hitherto this has been regarded as altogether indefinite, and it is true that the specimens are often in great confluent masses or sheets, the latter sometimes distorted by the lateral pressure which the limestone has experienced. The specimen from Tudor, however, figured by Sir W. E. Logan in the *Quarterly Journal* of the Geological Society, 1867, p. 253, and that described by me in the "Proceedings of the

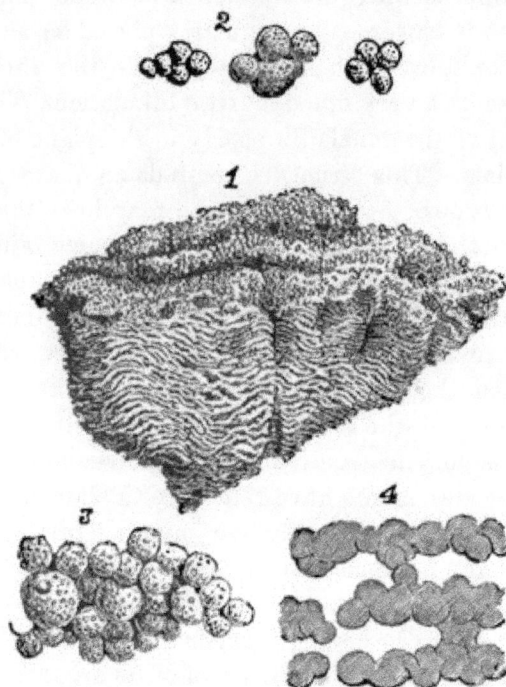

Fig. 10. *Eozoon Canadense.* (1) Small specimen disengaged by weathering. (2) Acervuline cells of upper part—magnified. (3) Tuberculated surface of lamina—mag. (4) Laminæ of Serpentine in section, representing casts of the sarcode—mag.

American Association" in 1876, and figured in my work, "Life's Dawn on Earth," gave the idea of a turbinate form more or less broad. More recently additional specimens weathered out of the limestone of Côte St. Pierre have been

obtained by Mr. E. H. Hamilton, who collected for me at
that place; and these, on comparison with several less perfect specimens in our collections, have established the fact
that the normal shape of young and isolated specimens of
Eozoön Canadense is a broadly-turbinate, funnel-shaped, or
top-shaped form, sometimes with a depression on the upper
surface giving it the appearance of the ordinary cup-
shaped Mediterranean sponges. (Fig. 11.) These specimens also show that there is no theca or outer coat either
above or below, and that the laminæ pass outwards without change to the margin of the form, where, however, they
tend to coalesce by subdividing and bending together. The
laminæ are thickest at the base of the inverted cone, and
become thinner and closer on ascending, and at the top they
become confounded in a general vesicular or acervuline
layer. I feel now convinced that broken fragments of this
upper surface scattered over the sea-bottom formed those
layers of *Archæosphærinæ* which at one time I regarded as
distinct organisms.

It is to be observed, however, that other forms of Eozoön
occur. More especially there are rounded or dome-shaped
masses, that seem to have grown on ridges or protuber-
ances, now usually represented by nuclei of pyroxene.

(2.) *Osculiform tubes.*

In the large number of specimens of Eozoön which have
been cut or sliced in various directions, and are now in our
museum at Montreal, it has become apparent that there are
more or less cylindrical depressions or tubes, sometimes
filled with serpentine and sometimes with inorganic calcite,
crossing the laminæ at right angles. These seem to occur
chiefly in the large and confluent masses, and are without
any regular or definite arrangement. In some of the nar-
rower openings of this kind the laminæ can be observed to
subdivide and become confluent on the sides of these tubes,
in the same manner as at the external surface. This circumstance induces me to believe that these are not accidental, but original parts of the structure, and intended to

admit water into the lower parts of the masses. (See Frontispiece.) A central canal of a similar kind is well shown in the accompanying illustration.

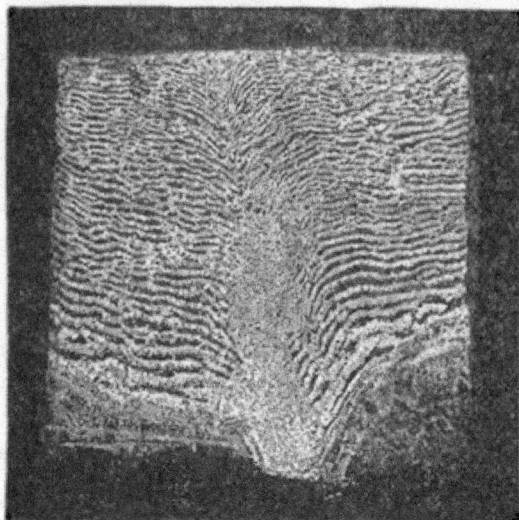

Fig. 12. Section of the base of a specimen of *Eozoon*. This specimen shows an osculiform, cylindrical perforation, cut in such a manner as to show its *reticulated wall* and the descent of the laminæ toward it. Two-thirds of natural size. From a photograph. Coll. Carpenter, also in Redpath Museum. [This illustration (from Prof. Prestwich's "Geology," vol. ii., p. 21) has been courteously lent by the Clarendon Press, Oxford.]

(3.) *Beds of Fragmental Eozoön.*

If Eozoön was an organism growing on the sea-bottom, it would be inevitable that it would be likely to be broken up, and in this condition to constitute a calcareous sand or gravel. I have already in previous pages noticed Laurentian limestones containing such fragments, from the Grenville band at Côte St. Pierre, from the Adirondack Mountains in New York State, from Chelmsford, Massachusetts, and from St. John, New Brunswick, as well as from Brazil and the Swiss Alps. Indeed, the Laurentian limestones of most parts of the world hold fragmental Eozoön. In the Peter Redpath Museum are some large slabs of Laurentian limestone sawn under the direction of Sir W. E. Logan, and showing irregular layers and detached masses

of Eozoön with layers or bands of limestone and of ophiolite. These are evidently layers successively deposited,
though somewhat distorted by subsequent movements. On
selecting specimens from the white and more purely calcareous layers, I was pleased to find that they abound in
fragments of laminæ of Eozoön, having the canals filled
either with dolomite or with colourless serpentine. Other
portions of the limestone show the peculiar granulated
structure characteristic of the calcareous laminæ of Eozoön,
but without any appearance of canals, which may in this
case be occupied with calcite, not distinguishable from the
substance of the laminæ. There are also indications in
these beds of limestones of the presence of Eozoön not infiltrated with serpentine, but having its laminæ either compressed together, or with the spaces between them filled
with calcite. There are other fragments which, from their
minute structure, I believe to be organic, but which are
apparently different from Eozoön.

(4.) *Veins of Chrysotile.*

I have in previous pages noticed the fact that the
veins of fibrous chrysotile which abound in serpentinous
limestones of the Laurentian are of secondary aqueous
origin, as they fill cracks or fissures not merely crossing
the beds of the limestone, but passing through the masses
of Eozoön and the serpentinous concretions which occur in
the beds. They must, therefore, have been formed by
aqueous action long after the deposition, and in some cases
after the folding and crumpling of the beds. In this
respect they differ entirely from the laminæ of Eozoön,
which have been subject to the same compression and folding with the beds themselves.

The chrysotile veins have, of course, no connection with
the structures of Eozoön, though they have often been mistaken for its more finely tubulated portion. With respect
to this latter, I believe that some wrong impressions have
been created by defining it too rigorously as a " proper

wall." In so far as I can ascertain, it consisted of finely divided tubes similar to those of the canal system, and composed of its finer subdivisions placed close together, so as to become approximately parallel. (See Fig. 4 above.)

(5.) *Nodules of Serpentine.*

Reference has been made in previous papers to the nodules and grains of serpentine found in the Eozoön limestone, but destitute of any structure. These nodules, as exhibited in the large slabs already referred to, have however often patches of Eozoön attached to or imbedded in them, and they appear to indicate a superabundance of this siliceous material accumulating by concretionary action around or attached to any foreign body, just as occurs with the flints in chalk. The layers and grains of serpentine parallel to the bedding appear to be of similar origin.

(6.) *State of Preservation.*

Recent observations more and more indicate the importance and frequency of dolomite as a filling of the canals, and also the fact that the serpentine deposited in and around the specimens of Eozoon is of various qualities. Dr. Sterry Hunt has shown that the purely aqueous serpentine found in the Laurentian limestones is of different composition from that occurring with igneous rocks, or as a product of the hydration of olivine. There are, however, different varieties even of this aqueous serpentine, ranging in colour from deep green to white; and one of the lighter varieties has the property of weathering to a rusty colour, owing to the oxidation of its iron. These different varieties of serpentine will, it is hoped, soon be analysed, so as to ascertain their precise composition. The mineral pyroxene, of the white or colourless variety, is a frequent associate of Eozoon, occurring often in the lower layers and filling some of the canals. Sometimes the calcareous laminæ themselves are partially replaced by a flocculent serpentine, or by pyroxenic grains imbedded in calcite.

(7.) *Other Laurentian Organisms.*

In a collection recently acquired by the Peter Redpath Museum, from the Laurentian of the Ottawa district, are some remarkable cylindrical or elongated conical bodies, from one to two inches in diameter, which seem to have occurred in connection with beds or nodules of apatite. They are composed of an outer thick cylinder of granular dark-coloured pyroxene, with a core or nucleus of white felspar; and they show no structure, except that the outer cylinder is sometimes marked with radiating rusty bands, indicating the decay of radiating plates of pyrite. They may possibly have been organisms of the nature of *Archæocyathus*; but such reference must be merely conjectural.

(8.) *Cryptozoum.*

The discovery by Prof. Hall, in the Potsdam formation of New York, and by Prof. Winchell in that of Minnesota, of the large laminated forms which have been described under the above name, has some interest in connection with Eozoon. I have found fragments of these bodies in conglomerates of the Quebec group, associated with Middle Cambrian fossils; and, whatever their zoological relations, it is evident that they occur in the Cambrian rocks under the same conditions as Eozoon in the Laurentian. I find also in the Laurentian limestones certain laminated forms usually referred to Eozoon, but which have thin continuous laminæ, with spongy porous matter intervening, in the manner of *Cryptozoum* or of *Loftusia*. Whether these are merely Eozoon in a peculiar state of preservation or a distinct structure, I cannot at present determine.

(9.) *Continuity and Character of containing Deposits.*

At a time when so many extravagant statements are made, more especially by some German petrologists, respecting the older crystalline rocks, it may be proper to state that all my recent investigations of the part of system

3

which I have called Middle Laurentian, especially in the district east of the Ottawa, vindicate the results of the late Sir William Logan as to the continuity of the great limestones, their regular interstratification with the gneisses, quartzose gneisses, quartzites, and micaceous schists, and their association with bedded deposits of magnetite and graphite, and also the regularity and distinctly stratified character of all these rocks. Farther, I regard the Upper Laurentian, independently of the great masses of Labradorite rock, which may be intrusive, as an important aqueous formation, characterised by peculiar rocks, more especially the anorthite gneisses. I am also of opinion that some of the crystalline rocks of the country west of Lake Superior are stratigraphically, and to a great extent lithologically, equivalent to the Upper Laurentian of St. Jerome and other places in the Province of Quebec, differing chiefly in the greater or less abundance of intrusive igneous rocks.

(10.) *Imitative Forms.*

The extraordinary mistakes made by some lithologists in studying imperfect examples of Eozoon and rocks supposed to resemble it, and which have gained a large amount of currency, have rendered necessary the collection and study of a variety of laminated rocks, and considerable collections of these have been made for the Peter Redpath Museum. They include banded varieties of dolerite and diorite, of gneiss, of apatite and of tourmaline with quartz, laminated limestone with serpentine, graphic granites, and a variety of other laminated and banded materials, which only require comparison with the genuine specimens to show their distinctness, but many of which have nevertheless been collected as specimens of Eozoon. I do not propose to enter into any detailed description of these here, but may hope, with the aid of Dr. Harrington, to notice them in forthcoming Memoirs of Peter Redpath Museum.

It is easy for inexperienced observers to mistake laminated concretions and laminated rocks either for *Stromatopora*

or for *Eozoon*, and such misapprehensions are not of infrequent occurrence. As to concretions, it is only necessary
to say that these, when they show concentric layers, are
deficient altogether in the primary requirements of laminæ
and interspaces; and under the microscope their structures
are either merely fragmental, as in ordinary argillaceous
and calcareous concretions, or they have radiating crystalline fibres like oolitic grains. Laminated rocks, on the
other hand, present alternate layers of different mineral
substances, but are destitute of minute structures, and are
either parallel to the bedding or to the planes of dykes and
igneous masses. In the Montreal mountain there are
beautiful examples of a banded dolerite in alternate layers
of black pyroxene and white felspar. These occur at the
junction of the dolerite with the Silurian limestone through
which it has been erupted. Laminated gneissose beds also
abound in the Laurentian. Still more remarkable examples
are afforded by altered rocks having thin calcite bands,
whether arising from deposition or from vein-segregation.
One of these now before me is a specimen from the collection
of Dr. Newberry, and obtained at Gouverneur, St. Lawrence
County, New York. It presents thick bands of a peculiar
granitoid rock containing highly crystalline felspar and
mica with grains of serpentine; these bands are almost a
quarter of an inch in thickness, and are separated by interrupted parallel bands of calcite much thinner than the
others. The whole resembles a magnified specimen of
Eozoon, except in the absence of the connecting chamber-
walls and of the characteristic structures. A similar rock
has been obtained by Mr. Vennor on the Gatineau; but it
is less coarse in texture though equally crystalline, and
appears to contain hornblende and pyroxene. These are
both Laurentian, and I consider it not impossible that they
may have been organic; but they lack the evidence of
minute structure, and differ in important details from
Eozoon. Another specimen from the Horseshoe Mountain
in the Western States (I regret that I have mislaid the
name of the gentleman to whom I am indebted for this

specimen) is a limestone with perfectly regular and uniform
layers of minute rhombohedral crystals of dolomite. The
layers vary in distance regularly in the thickness of the
specimen from two millimetres to one, and must have resulted
from the alternate deposition in a very regular manner of
dolomite and limestone. These are but a few of the examples
of imitative structures which might readily be confounded
with *Eozoon*, or which, if resulting from organic growth,
have lost all decisive evidence of the fact.

Perhaps still more puzzling imitative forms are those
referred to by Hahn, which occur in some felspars, and
which I have found in great beauty in certain crystals of
orthoclase from Vermont. They are ramifying tubes
resembling the canal-system of *Eozoon*, and are evidently a
peculiar form of gas-cavities or inclusions. Similar appearances are, however, often presented by the more minute
and microscopic varieties of graphic granite, in which the
little plates might readily be mistaken, in certain sections,
for organic tubulation.

In the present state of knowledge, it is perhaps more
excusable to mistake such things for organic structures than
to deny the existence of true organic structures because
they resemble such forms. Those who have examined
moss-gates are familiar with the fact that while some show
merely crystals of peroxide of iron or oxide of manganese,
others present the forms of *Vaucheriæ* or *Confervæ*. So if
one were to place side by side some fibres of asbestos,
spicules of *Tethea*, and coniferous wood, preserved, like
some from Colorado, as separate white siliceous fibres, they
might appear alike; but, even if thoroughly mixed together,
the microscope should be able easily to distinguish them.
I have specimens of fossil wood, collected by Hartt in Brazil,
which have been mineralized by limonite in such a manner
that no one, without microscopic examination, could believe
them to be other than fibrous brown hæmatite. Such
difficulties the micro-geologist must expect to find, and by
patient observation to overcome.

(11.) *Alternation of Mineral Layers.*

It has been suggested by Mr. Julien* and others that Eozoönal structure may be due to the alternation of mineral layers formed in the passage-beds between concretions and other mineral masses, and their enclosing matrix. The objections to this view are :

1. Laminated passage-rocks and laminated concretionary forms have only simple laminæ, whereas *Eozoon* has connected or reticulating laminæ.

2. Laminated passage-rocks have no structure other than crystalline. *Eozoon* has beautiful tubulation in its calcareous walls, besides large tubes or oscula.

3. Sometimes (not usually) pyroxene is the siliceous part of *Eozoon ;* or, as we hold, the mineralizing agent. More usually it is serpentine, sometimes loganite, or dolomite, or mere earthy limestone. It is not possible that all these minerals should assume the same forms.

4 Pyroxene and serpentine both occur in nodules and bands in the Laurentian limestones, and in most cases without any traces of *Eozoon*, while *Eozoon* occurs in the limestone remote from such nodules and bands, where no passage of any kind can occur, and presents distinct forms.

5. There are only two localities known to me, one in a quarry near Côte St. Pierre, and one at Burgess, where a bed with badly preserved *Eozoon* occurs in a manner which would even suggest such an idea. Pyroxene is present in the one case, and loganite in the other.

6. I have often thought of this suggested explanation, and have compared *Eozoon* with all sorts of banded and passage-rocks taken from the Laurentian and other formations, but have seen no reason to adopt such a view for *Eozoon*. I have accumulated in the Peter Redpath Museum at Montreal as above stated, a very large number of laminated and passage-rocks and concretions for purposes of comparison.

7. How on such an hyphthesis can we explain the beds

* Proceed. Amer. Assoc. vol. xxxiii. 1884, pp. 415, 416.

of limestone composed of or filled with fragments of *Eozoon*?

8. I append to this section a few extracts from previous articles on the subject, which bring up important points into the details of which it has been impossible to enter in the above remarks.

1. The complicated theory of pseudomorphism and replacement, advocated by Messrs. Rowney and King, may be dismissed at once. Independently of the insuperable chemical difficulties which have been pointed out by Dr. Hunt*, and which he has discusssed very fully in his "Papers on Chemical Geology," we have the further facts that no replacement of serpentine by calcite is indicated by the relations of these minerals to each other, while such replacement as does occur is in the other direction, or the change of calcite into serpentine, as evidenced by the state of preservation of some specimens of *Eozoon*, above referred to. Further, this theory fails to give any explanation of the specimens mineralized by pyroxene, dolomite, and calcite, or to account for the nummuline wall, except by attributing it to the alteration of chrysotile, which is inadmissible, as the veins of this mineral are newer than the walls supposed to have been derived from them.

2. Inasmuch as many apparently concretionary grains and lenticular masses of serpentine exist in the Laurentian limestones, it may be supposed possible that *Eozoon* is merely a modification of these concretionary forms. In this case, the filling of each lamina and chamberlet of *Eozoon* must be regarded as a separate concretion; and even if we could suppose some special cause to give regularity and uniformity to such concretions in some places and not in others, we still have unaccounted for the canals and tubuli, or the delicate threads of serpentine representing them. Further, we have to suppose that a tendency to this regular and complicated arrangement has affected in the same way minerals so diverse as serpentine, loganite, pyroxene, and dolomite.

3. The only remaining theory is that of infiltration by serpentine of cavities previously existing in the calcite. There is no chemical objection to this, inasmuch as we know of the infiltration of fossils, in other formations by minerals akin to serpentine; and in these limestones the veins of fibrous serpentine have evidently been introduced by aqueous action subsequently to the production or fossilization of the *Eozoon*. Further, the white pyroxene of the Laurentian limestones, and the loganite and dolomite, are all known

---

* Quart. Journ. Geol. Soc. vol. **xxiii**. p. 260.

to have been produced by aqueous deposition. The only question remaining is, Whence came the original calcite skeleton with laminæ, chambers, canals, and tubuli to be so infiltrated? The answer is given in the comparison with the tests of Foraminifera, originally proposed by the writer, and illustrated in so conclusive a manner by the researches of Dr. Carpenter.

I may add, in conclusion, that had geologists generally the opportunity of studying *Eozoon in situ*, in good exposures, like that at St. Pierre, they would much more fully understand and appreciate the arguments for its organic nature, than when they have had opportunities of examining only polished specimens and slices†. Its stromatoporoid masses, projecting from the weathered beds of limestone, would at once attract the attention of any collector; and the whole conditions of its occurrence, whether entire or in fragments, are precisely those of fossil corals in the Silurian limestones. Further, the symmetry and uniformity of its habit of growth are much more apparent when they can be studied in large specimens prepared by natural weathering or by treatment with an acid.

V. NOTES ON CERTAIN PECULIAR SPECIMENS.

1. SPECIMENS FROM CÔTE ST. PIERRE.

In order to test the state of preservation of the canal system and nummuline layer, I treated a great number of specimens from different parts of the bed with a dilute acid. The result was, that in all, the canal system could be detected in greater or less perfection in the thicker laminæ near the base of the forms, and in some through a great number of the laminæ. The structure of the nummuline layer was not so constantly preserved, its tubuli not being infiltrated in some parts, so that it appears as a structureless band, or fails altogether to be visible. In no instance could it be seen to pass into chrysotile, as recently affirmed by Messrs. Rowney and King*, although chrysotile veins often run very near to or across the walls. The nummuline layer is almost always distinctly limited by parallel surfaces, with its tubes at right angles, or nearly so, to these.

---

* Trans. Royal Irish Academy, 1871.

† I have been sorry to find, from specimens in the cabinets of my friends, that some dealers are in the habit of circulating specimens labelled "*Eozoon Canadense*" which have no trace of the structures of the fossil, but are either badly preserved acervuline portions or merely ordinary serpentine marble. Such specimens can, of course, only mislead, and may produce much unnecessary scepticism.

* Ann. & Mag. of Nat. Hist. May 1874.

By careful scrutiny of the beds we were enabled to detect two new forms of *Eozoon*, which may eventually prove to be distinct species, but which for the present may be regarded as varietal forms.

One of these, found *in situ* by Mr. Weston, is flat in form and very finely laminated, with thin walls, except near the lower part. where there is some supplemental skeleton with finer and more curved canals than usual.   The thin walls or laminæ of the ordinary skeleton are connected by very frequent vertical pillars or partitions, and are as numerous as thirty in half an inch, while the whole thickness of the specimen does not exceed an inch.   It thus very closely resembles some of the Devonian and Silurian *Stromatoporæ*, especially when seen on the weathered surface.   It may be named in the meantime variety *minor*.

The second occurs in more or less oval patches a few inches in diameter, limited by a sort of a frame or border of compact serpentine, and presenting under the microscope an aggregation of small acervuline chamberlets, having the proper wall filled with unusually long parallel tubes, and with little development of supplemental or intermediate skeleton.   The appearance of parallel tubulation running through and past several successive chamberlets was more conspicuous in these specimens than in the ordinary acervuline *Eozoon*, and the chamberlets themselves more cylindrical and tortuous.   These specimens may either be portions of the acervuline superficial part of *Eozoon* broken off and separately preserved, or they may constitute a distinct varietal form.   As the latter seems on the whole most probable, I would name this form variety *acervulina* (Fig. 13)

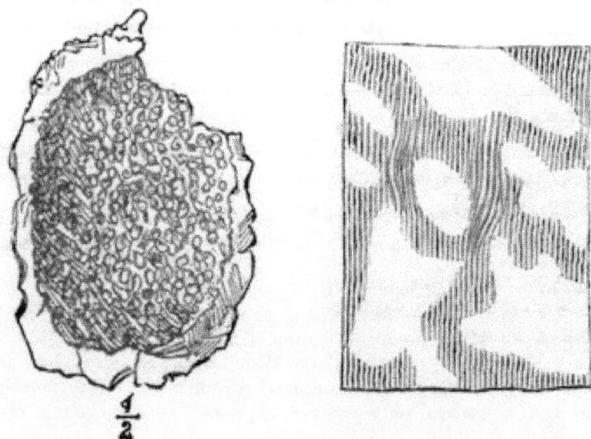

ig. 13.   *Eozoon*, peculiar variety—half natural size ; and tubulation magnified ôte St. Pierre.

These varieties are of much more rare occurrence than the ordinary type of *Eozoon*.

The ordinary specimens of *Eozoon* found at St. Pierre are mineralized with serpentine ; but fragments imbedded in the dolomitic limestones have 'their canals filled with a transparent mineral which, from i:s optical character, is evidently dolomite, though the quantity obtained was not sufficient for any definite chemical test. Parts of the canals in these specimens were filled with calcite, as shown by its dissolving entirely away in a dilute acid. In one of the serpentinous specimens also I have observed that, while portions of the groups of canals, especially the basal portions, are filled with serpentine, the extremities of the canals and their finer branches present, under polarized light, the aspect of calcite; and that they are filled with this mineral is proved by these portions of the canal-filling been entirely removed when treated with dilute acid. It would thus appear that in these specimens, while the terminal parts of the canals have been filled with calcite, the basal portions have been occupied by serpentine. This is not, however, a new fact, as similar appearances have been already described both by Dr. Carpenter and the writer.

In one specimen I observed a portion of the fossil entirely replaced by serpentine, the walls of the skeleton being represented by a lighter-coloured serpentine than that filling the chambers, and still [retaining traces of the canals. The walls thus replaced by serpentine could be clearly traced into connexion with the portions of those still existing as calcite. This shows that the serpentine, like the quartz in silicified shells and corals, has had the power of replacing the calcite of the fossils; and I believe that its partial action in this way accounts for some irregularities observed in the less perfectly preserved specimens. Nor is it improbable, as Dr. Hunt has already suggested, that some of the masses of serpentine and pyroxene on which specimens of *Eozoon* are based, may represent older and more perfectly mineralized masses of the fossil.

In some of the specimens of *Eozoon*, the superficial laminæ are apparently broken and displaced in such a manner as to suggest the idea that partial disintegration by the waves had taken place before they were finally buried. It is also observable that in some of the masses the compression to which they have been subjected has produced a microscopic faulting, which slightly displaces the laminæ.

2. SPECIMEN OF EOZOON FROM TUDOR, ONTARIO.[*]

This very interesting specimen, discovered by Mr. Vennor of the Canadian Geological Society, in the slightly altered carbonaceous limestone of the Hastings group, was submitted to me by Sir W. E. Logan. It is, in my opinion, of great importance, as furnishing a conclusive answer to all those objections to the organic nature of *Eozoon* which have been founded on comparisons of its structures with the forms of fibrous, dendritic, on concretionary minerals,— objections which, however plausible in the case of highly crystalline rocks, in which organic remains may be simulated by merely mineral appearances readily confounded with them, are wholly inapplicable to the present specimen.

1. GENERAL APPEARANCE.—The fossil is of a clavate form, six and a half inches in length, and about four inches broad. It is contained in a slab of dark-coloured, coarse, laminated limestone, holding sand, scales of mica, and minute grains and fibres of carbonaceous matter. The surface of the slab shows a weathered section of the fossil, and the thickness remaining in the matrix is scarcely two lines, at least in the part exposed. The septa, or plates of the fossil, are in the state of white carbonate of lime, which shows their form and arrangement very distinctly, in contrast to the dark stone filling the chambers. The specimen lies flat in the plane of stratification, and has probably suffered some compression. Its septa are convex towards the broad end, and somewhat undulating. In some places they are continuous half-way across the specimen; in other places they divide and re-unite at short distances. A few transverse plates, or connecting columns, are visible; and there are also a number of small veins or cracks passing nearly at right angles to the septa, and filled with carbonate of lime, similar in general appearance to the septa themselves.

On one side, the outline of the fossil is well preserved. The narrow end, which I regard as the basal portion, is rounded. The outline of the side first bends inward, and then outward, forming a graceful double curve, which extends along the greater part of the length. Above this is an abrupt projection, and then a sudden narrowing: and in the middle of the narrow portion, a part has the chambers obliterated by a white patch of carbonate of lime, below which some of the septa are bent downward in the middle. This is probably an effect of mechanical injury, or of the interference of a calc-spar vein.

---

[*] This specimen is in the Museum of the Geological Survey at Ottawa. It is represented by a photograph in the Redpath Museum. Its description was originally published in the Journal of the Geological Society of London, August, 1867.

With the exception of the upper part above referred to, the septa are seen to curve downward rapidly toward the margin, and to coalesce into a lateral wall, which forms the defined edge or limit of the fossil, and in which there are some indications of lateral orifices opening into the chambers. It is worthy of remark that, in this respect, the present specimen corresponds exactly with that which was originally figured by Sir W. E. Logan in the 'Geology of Canada,' p. 49, and which up to that time was the only other specimen that exhibited the lateral limit of the form.

On the side next the matrix, the septa terminate in blunt edges, and do not coalesce; as if the organism had been attached by that surface, or had been broken before being imbedded.

MICROSCOPIC CHARACTERS.—Under the microscope, with a low power, the margins of the septa appear uneven, as if eroded or tending to an acervuline mode of growth; but occasionally they show a distinct and regular margin. For the most part merely traces of structure are preserved, consisting of small parts of canals, filled with the dark colouring-matter of the limestone. In a few places, however, these appear as distinct bundles, similar to those in the Grenville specimens, but of finer texture.

In a few rare instances only can I detect, with a higher power, in the margin of some of the septa, traces of the fine tubulation characteristic of the proper chamber-wall of *Eozoon.* For the most part this seems to have been obliterated by the infiltration of the tubuli with colourless carbonate of lime, similar to that of the skeleton.

In comparing the structure of this specimen with that of those found elsewhere, it would appear that the chambers are more continuous, and wider in proportion to the thickness of the septa, and that the canal-system is more delicate and indistinct than usual In the two former respects the specimens from the Calumet and from Burgess approach that now under consideration more nearly than do those from Grenville and Petite Nation: but it would be easy, even in the latter, to find occasional instances of a proportion of parts similar to that in the present example. General form is of little value as a character in such organisms; and so far as can be ascertained, this may have been the same in the present specimen and in that originally obtained from the Calumet, while in the specimens from Grenville a massive and aggregative mode of growth seems to have obliterated all distinctness of individual shape. Without additional specimens, and in the case of creatures so variable as the Foraminifera, it would be rash to decide whether the differences above noticed are of specific value, or depend on age, variability, or state of preservation. For this reason I refer the specimen for the present to *Eozoon Canadense,* merely distinguishing it as the Tudor variety.

From the state of preservation of the fossil, there are no crystalline structures present which can mislead any ordinary skilful microscopist, except the minute veins of calcareous spar traversing the septa, and the cleavage-planes which have been developed in the latter.[*]

SPECIMEN FROM MADOC.—In a letter to Dr. Carpenter, quoted by him in the 'Quarterly Journal of the Geological Society' for August, 1866, p. 228, I referred to the occurrence of *Eozoon* preserved simply in carbonate of lime. The specimens which enabled me to make that statement were obtained at Madoc, near Tudor, this region being one in which the Laurentian rocks of Canada appear to be less highly metamorphosed than is usual.[†] The specimens from Madoc, however, were mere fragments, imbedded in the limetone, and incapable of showing the general form. I may explain, in reference to this, that long practice in the examination of these limestones has enabled me to detect the smallest fragments of *Eozoon* when present, and that in this way I had ascertained the existance of this fossil in one of the limestones of Madoc before the discovery of the fine specimen from Tudor.

This fragment projected from the surface of the limestone, being composed of a yellowish dolomite, and looking like a fragment of a thick shell.   When sliced, it presents interiorly a crystalline dolomite, limited and separated from the enclosing rock by a thin wall having a granular or porous structure and excavated into rounded recesses in the manner of *Eozoon*.   It lies obliquely to the bedding, and evidently represents a hollow flattened calcareous wall filled by infiltration.   The limestone which afforded this form was near the beds holding the worm-burrows described in the Journal of the Geological Society for November, 1866.

The following are remarks by Dr. Carpenter on the Madoc specimen;—"A thin section of this body, carefully examined microscopically, presents numerous and very characteristic examples of the canal system of *Eozoon*, exhibiting both the large widely branching systems of canals and the smaller and more penicillate tufts shown in the most perfect of the serpentinous specimens—but with the difference, that the canals, being filled with a material either identical with or very similar to that of the substance in which they are excavated, are so transparent as only to be brought into view by careful management of the light."

---

[*] I cannot, after examination of the specimen, and of others subsequently obtained by Sir W. E. Logan, attach any value to the supposition of Messrs. Rowney & King that the Tudor specimen has been produced by infiltration of carbonate of lime into veins. The mechanical arrangement of the laminæ and their microscopic structure forbid such a supposition, as well as the comparison of them with actual calcareous veins occurring in the same rock.

[†] They may be later than Laurentian, possibly Huronian.

3. SPECIMENS FROM LONG LAKE AND WENTWORTH.

Specimens from Long Lake, in the collection of the Geological Survey of Canada, exhibit white crystalline limestone with light-green compact or septariiform* serpentine, and much resemble some of the serpentine-limestones of Grenville. Under the microscope the calcareous matter presents a delicate areolated appearance, without lamination; but it is not an example of acervuline *Eozoon*, but rather of fragments of such a structure, confusedly aggregated together, and having the interstices and cell-cavities filled with serpentine. I have not found in any of these fragments a canal-system similar to that of *Eozoon Canadense*, though there are casts of large stolons, and, under a high power, the calcareous matter shows in many places the peculiar granular or cellular appearance which is one of the characters of the supplemental skeleton of that species. In a few places a tubulated cell-wall is preserved, with structure similar to that of *Eozoon Canadense*.

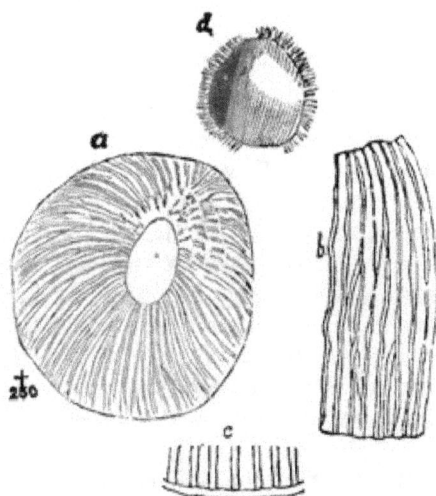

Fig. 14. Chamberlets from Long Lake. (*a*) Entire form showing tubulation; (*bc*) same more magnified; (*d*) cast of interior with casts of tubuli.

Specimens of Laurentian limestone from Wentworth, in the collection of the Geological Survey, exhibit many rounded siliceous bodies, some of which are apparently grains of sand, or small pebbles; but others, especially when freed from calcareous matter by a dilute acid, appear as rounded bodies, with rough surfaces, either separate or aggregated in lines or groups, and having minute

---

* I use the term ' septariiform ' to denote the *curdled* appearance so often presented by the Laurentian serpentine.

vermicular processes projecting from their surfaces. (Fig. 14.) At first sight these suggest the idea of spicules ; but I think it on the whole more likely that they are casts of cavities and tubes belonging to some calcareous Foraminiferal organism which has disappeared. Similar bodies, found in the limestone of Bavaria, have been described by Gümbel, who interprets them in the same way.* They may be also compared with the silicious bodies mentioned in a former paper as occurring in the Loganite filling the chambers of specimens of *Eozoon* from Burgess.

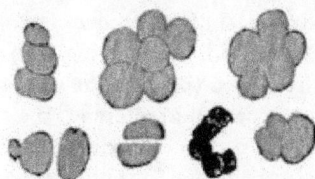

Fig. 14.   Cast of Chamberlets ; cavities of Eozoon, Burgess.

4. CANALS NOT TO BE CONFOUNDED WITH DENDRITIC MINERALS, &C.

The discovery of the specimen from Tudor, above described, may appear to render unnecessary any reference to the elaborate attempt made by Profs. King and Rowney to explain the structures of *Eozoon* by a comparison with the forms of fibrous and dendritic minerals,† more especially as Dr. Carpenter has already shown their inaccuracy in many important points. I think, however, that it may serve a useful purpose shortly to point out the more essential respects in which this comparison fails with regard to the Canadian specimens—with the view of relieving the discussion from matters irrelevant to it, and of fixing more exactly the limits of crystalline and organic forms in the serpentine-limestones and similar rocks.

The fundamental error of the writers above named arises from defective observation—in failing to distinguish, in the Canadian limestones themselves, between organic and crystalline forms. This is naturally followed by the identification of all these forms, whether mineral or organic, with a variety of purely crystalline arrangements occurring in other rocks, leading to their attaching the term "Eozoonal" to any rock which shows any of the characters, whether mineral or organic, thus arbitrarily attached to the Canadian *Eozoon*. This is obviously a process by which the structure of any fossil might be proved to be a mere *lusus naturæ*.

* Proceedings of Royal Academy of Munich, 1866 ; Q.J.G.S vol xxii. pt. i. p. 185 *et seq.* ; also, Can. Naturalist, vol. iii. p. 81.
  † Quart. Journ. Geol. Soc. vol. xxii. pt. ii. p. 23.

A notable illustration of this is afforded by their regarding the veins of fibrous serpentine, or chrysotile, which occur in the Canadian specimens, as identical with the tubulated cell-wall of *Eozoon* —although they admit that these veins traverse all the structures indifferently and do not conform to the walls of the chambers. But any microscopist who possesses specimens of *Eozoon* containing these chrysotile veins may readily satisfy himself that, under a high power, they resolve themselves into *prismatic crystals in immediate contact with each other;* whereas, under a similar power, the true cell-wall is seen to consist of *slender, undulating, rounded threads of serpentine, penetrating a matrix of carbonate of lime.* Under polarized light, more especially, the difference is conspicuously apparent.

It would also appear that the radiating and sheaf-like bundles of crystals of tremolite, or similar prismatic minerals, which occur in the Canadian serpentines, and also abound in those of Connemara, have been confounded with the tubulation of *Eozoon*; but these crystals have no definite relation to the forms of that fossil, and and often occur where these are entirely absent; and in any case they are distinguished by their straight prismatic shape and their angular divergence from each other. Much use has also been made of the amorphous masses of opaque serpentinous matter which appear in some parts of the structure of *Eozoon*. These I regard as, in most cases, simply results of alteration or defective preservation, though they might also arise from the presence of foreign matters in the chambers, or from an incrustation of mineral matter before final filling up of the cells. Generally their forms are purely inorganic; but in some cases they retain indications of the structures of *Eozoon*.

With reference to the canal-system of *Eozoon*, no value can be attached to loose comparisons of a structure so definite with the forms of dendritic silver and the filaments of moss-agates; still less can any resemblance be established between the canal system and vermicular crystals of mica. These occur abundantly in some serpentines from the Calumet, and might readily be mistaken for organic forms; but their rhombic or hexagonal outline when seen in cross section, their transverse cleavage planes, and their want of any definite arrangement or relation to any general organic form, are sufficient to undeceive any practised observer. I have not seen specimens of the metaxite from Reichenstein referred to by Messrs. King and Rowney; but it is evident, from the description and figure given of it, that, whether organic or otherwise, it is not similar to the canals of *Eozoon Canadense*. But all these and similar comparisons are evidently worthless when it is considered that they have to account for definite, ramifying, cylindrical forms, penetrating a skeleton or matrix of limestone, which has itself a definite arrangement and structure, and, further, when we find that these forms are represented by substances so diverse as serpentine, pyroxene, limestone, and carbonaceous matter. This is intelligible on the supposition of tubes filled with foreign matters, but not on that of dendritic crystallization.

If all specimens of *Eozoon* were of the acervuline character, the comparisons of the chamber-casts with concretionary granules might have some plausibility. But it is to be observed that the laminated arrangement is the typical one; and the study of the larger specimens, cut under the direction of Sir W. E. Logan, shows that these laminated forms must have grown on certain strata-planes before deposition of the overlying beds, and that the beds are, in part, composed of the broken fragments of similar laminated structures. Further, much of the apparently acervuline *Eozoon* rock is composed of such broken fragments, the interstices between which should not be confounded with the chambers; while the fact that the serpentine fills such interstices as well as the chambers shows that its arrangement is not concretionary.*

It is also to be observed that examination of a number of limestones, other than Canadian, by Messrs. King and Rowney, has obliged them to admit that the laminated forms in combination with the canal system are "essentially Canadian," and that the only instances of structures clearly resembling the Canadian specimens are afforded by limestones Laurentian in age, and in some of which (as, for instance, in those of Bavaria and Scandinavin) Carpenter and Gümbel have actually found the structure of *Eozoon.* The other serpentine limestones examined (for example, that of Skye) are admitted to fail in essential points of structure; and the only serpentine believed to be of eruptive origin, examined by them, is confessedly destitute of all semblance of *Eozoon.* Similar results have been obtained by the more careful researches of Prof. Gümbel, whose paper is well deserving of study by all who have any doubts on this subject.

In the above remarks I have not referred to the disputed case of the Connemara limestones ; but I may state that I have not been able to satisfy myself of the occurrence of the structures of *Eozoon* in such specimens as I have had the opportunity to examine.* It is perhaps necessary to add that there exists in Canada abundance of Laurentian limestone which shows no indication of the

---

* I do not include here the "septariiform" structure referred to above, which is common in the Canadian serpentine and has no connexion with the forms of the chambers.

structures of *Eozoon*. In some cases it is evident that such structures have not been present. In other cases they may have been obliterated by processes of crystallization. As in the case of other fossils, it is only in certain beds, and in certain parts of those beds, that well-characterized specimens can be found. I may also repeat here that in the original examination of *Eozoon*, in the spring of 1864, I was furnished by Sir W. E. Logan with specimens of all these limestones, and also with serpentine limestones of Silurian age; but that, while all possible care was taken to compare these with the specimens of *Eozoon*, it was not thought necessary to publish notices of the crystalline and concretionary forms observed, many of which were very curious, and might afford materials for other papers of the nature of that criticised in the above remarks.

5. DR. CARPENTER ON CANALS FILLED WITH CARBONATE OF LIME.

" The examination of a large number of sections of a specimen of *Eozoon*, recently placed in my hands by Sir William Logan, in which the canal system is extraordinarily well preserved, enables me to supply a most unexpected confirmation of Dawson's statements in regard to the occurrence of dendritic and other forms of this system, which cannot be accounted for by the intrusion of any foreign mineral; for many parts of the calcareous lamellæ in these sections, which, when viewed by ordinary transmitted light, appear quite homogeneous and structureless, are found, when the light is reduced by Collin's "graduating diaphragm," to exhibit a most beautiful development of various forms of canal system (often resembling those of Dawson's Madoc specimen noticed above), which cross the cleavage planes of the shell-substance in every direction. Now these parts, when subjected to decalcification, show no trace of canal system ; so that it is obvious, both from their optical and from their chemical reactions, that the substance filling the canals must have been *carbonate of lime*, which has thus completely solidified the shell layer, having been deposited in the canals previously excavated in its interior, just as crystalline carbonate of lime fills up the reticular spaces of the skeleton of Echinodermata fossilized in a calcareous matrix. This fact affords conclusive evidence of *organic structure*, since no conceivable process of crystallization could give origin to dendritic extensions of carbonate of lime disposed on exactly the same crystalline system with the calcite which includes them, the two substance being

* Such Irish specimens of serpentine limestone as I have seen, appear much more highly crystalline than the beds in Canada which contain *Eozoon*.

4

mineralogically homogeneous, and only structurally distinguishable by the effect of their junction surfaces on the course of faint rays of light transmitted through them."

VI. REPLIES TO PROFESSOR KARL MOBIUS* AND OTHERS.

The most serious attempt to invalidate the animal nature of Eozoön has been that of the above-named zoologist;* which was therefore deemed worthy of special notice and reply in the *American Journal of Science*, in which a summary of Prof. Möbius' memoir had been given.†

Eozoon Canadense has, since the first announcement of its discovery by Logan in 1859, attracted much attention, and has been very thoroughly investigated and discussed, and it has secured some recognition as organic.   Still its claims are ever and anon disputed, and as fast as one opponent is disposed of, another appears. This is in great part due to the fact that so few scientific men are fully in a position to appreciate the evidence respecting it.   Geologists and mineralogists look upon it with suspicion, partly on account of the great age and crystalline structure of the rocks in which it occurs, partly because it is associated with the protean and disputed mineral Serpentine, which some regard as eruptive, some as metamorphic, some as pseudomorphic, while few have had enough experience to enable them to understand the difference between those serpentines which occur in limestones, and in such relations as to prove their contemporaneous deposition, and those which may have resulted from the hydration of olivine or similar changes.   Only a few also have learned that *Eozoon* is only sometimes associated with serpentine, but that it occurs also mineralized with loganite, pyroxene, dolomite, or even earthy limestone, though the serpentinous specimens have attracted the most attention, owing to their beauty and abundance in certain localities. The biologists on the other hand, even those who are somewhat familiar with foraminiferal organisms, are little acquainted with the appearance of these when mineralized with silicates, traversed with minute mineral veins, faulted, crushed and distorted, as is the case with most specimens of *Eozoon*.   Nor are they willing to admit the possibility that these ancient organisms may have pre-

* Der Bau des Eozoon Canadense, von Karl Möbius, Professor der Zoologie in Kiel. Palæontographica, Band xxv.
† March, 1870.

sented a much more generalized and less definite structure than their modern successors. Worse, perhaps, than all these, is the circumstance that dealers and injudicious amateurs have intervened, and have circulated specimens of *Eozoon*, in which the structure is too imperfectly preserved to admit of its recognition, or even mere fragments of serpentinous limestone, without any structure whatever. I have seen in the collections of dealers, and even in public museums, specimens labelled " *Eozoon Canadense* " which have as little claim to that designation as a chip of limestone has to be called a coral or a crinoid.

The memoir of Prof. Möbius affords illustrations of some of these difficuties in the study of *Eozoon*. Prof. Möbius is a zoologist, a good microscopist, fairly acquainted with modern foraminifera, and a conscientious observer; but he has had no means of knowing the geological relations and mode of occurrence of *Eozoon*, and he has had access merely to a limited number of specimens mineralized with serpentine. These he has elaborately studied, and has made careful drawings of portions of their structures, and has described these with some degree of accuracy; and his memoir has been profusely illustrated with figures on a large scale. This, and the fact of the memoir appearing where it does, convey the impression of an exhaustive study of the subject, and since the conclusion is adverse to the organic character of *Eozoon*, this paper may be expected, in the opinion of many not fully acquainted with the evidence, to be regarded as a final decision against its animal nature. Yet, however commendable the researches of Möbius may be, when viewed as the studies of a naturalist desirous of satisfying himself on the evidence of the material he may have at command, they furnish only another illustration of partial and imperfect investigation, quite unreliable as a verdict on the questions in hand. The following considerations will serve to indicate the weak points of the memoir:—

1. A number of errors and omissions arise from want of study of the fossil *in situ*, and from want of acquaintance with its various states of preservation. Trivial errors of this kind are his referring to my photograph in Plate III, of the " Dawn of Life," as if it were natural size, and his stating that the larger specimens have fifty laminæ, whereas they have often more than an hundred. More important is his failing to appreciate aright the occurrence of *Eozoon* in certain layers of regularly bedded limestones, the rounded or club-shaped forms of the more perfect specimens, the manner in which the layers become confluent at the edges of the forms, as described by Sir W. E. Logan and myself, or the amount of crushing and fracture which most of the specimens exhibit.

Thus he fails to convey any adequate idea of the Stromatoporoid forms and mode of occurrence of the organism, or indeed of its general character and probable mode of growth. Farther he treats it from the first as a mere laminated aggregate of calcite and serpentine, without reference to its occurrence in any other state, and also without reference to the fragmental limestones in part made up of its remains. He objects strongly to the want of definiteness of form and distribution in the chambers and connecting passages, without making allowance for defects of preservation, or mentioning the similar want of defined form in some *Stromatoporæ*. He admits, however, that the modern *Carpenteria* and its allies are in some respects equally indefinite. He farther objects to the impossibility of detecting regular primary chambers like those in modern foraminifera, but seems not to be aware that, as I have recently shown, some *Stromatoporæ* originate in a vesicular, irregular mass of cells, and that in *Loftusia*, both the Eocene *L. Persica*, and the Carboniferous *L. Columbiana*, the primary chamber is represented by a merely cancellated nucleus.*

2. With reference to the finely tubulated proper wall of Eozoon, he has fallen into an error scarcely excusable in an observer of his experience, except on the plea of insufficient access to specimens. He confounds the proper wall with the chrysotile veins traversing many of the specimens, and obviously more recent than the bodies whose fissures they fill. That he does so is apparent from his stating that the proper wall structure sometimes crosses the bands of serpentine and calcite, and also that it presents a series of parallel, four-sided prisms, whereas, when at all perfectly preserved, it shows a series of cylindrical threads penetrating a calcite wall. That some of his specimens have contained the proper wall fairly preserved is obvious from his own figures, in which it is possible to recognize both this structure and chrysotile veins, though confounded by him under the same designation. He objects somewhat naïvely, that many of the chambers fail to exhibit this nummuline wall, and that it sometimes presents a ragged appearance or is altogether opaque. In point of fact it can appear distinctly, either in decalcified specimens or in slices, only when the minute tubes are filled with some substance optically distinguishable from calcite, or not acted on by dilute acid. When the proper wall is merely calcareous (and I have specimens showing that it is often in this state, and without any serpentine in its pores), its structure is ordinarily invisible, and it is the same when the calcareous skeleton has from any cause lost its transparency or

* See Journal of London Geol. Soc., January, 1878.

1.

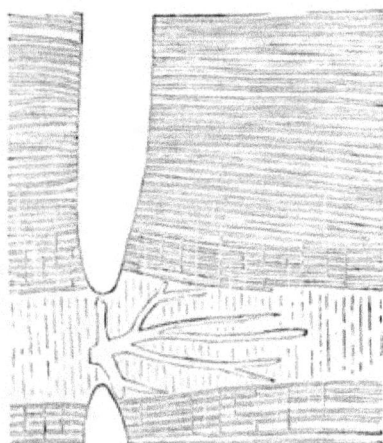

2.

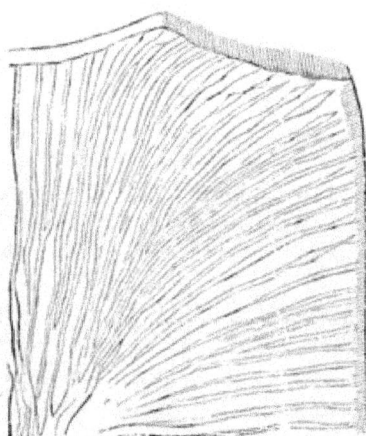

Fig. 15. Comparison of finer tubuli of *Nummulina* and *Eozoon* (after Möbius. (1) Tubuli and Canals of Nummulina ; (2) Tubuli and Canals of Eozoon.

has been replaced by some other mineral substance. Even in thickish slices, the tubes, though filled with serpentine, may be so piled on one another as to be indistinct. All this may be seen in Tertiary *Nummulites.* When wholly calcareous their tubulation is often quite invisible, and when imperfectly injected with glauconite or other silicates, they often present a very irregular appearance. If Professor Möbius will study the Nummulites injected with

glauconite from Kempfen,* Bavaria, in addition to the casts of
Polystomella from the Ægean to which he refers, he will be better
able to appreciate these points.   It may be worth repeating here
that, in examining the original specimens of Eozoon, I did not
recognize the proper wall.   I did not doubt that it must have
existed in some form, since I could easily detect the canals in the
supplemental skeleton ; but I did not wonder at its non-appearance,
knowing the chances against its preservation in a recognizable
form.   Its discovery was due to the subsequent investigations of
Dr. Carpenter. †

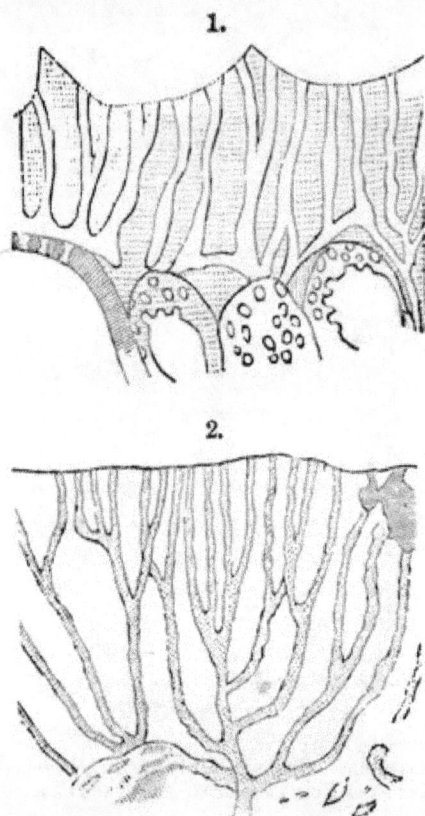

Fig. 16. Comparison of canals of *Calcarina* and *Eozoon* (after Möbius).   (1)
Canals of Calcarina ; (2) Canals of *Eozoon.*

* I am indebted to Dr. Otto Hahn for specimens of these most interesting
fossils.

† It may deserve mention here that the Carboniferous *Fusulina* very rarely
shows its tubulated wall, and that Dr. Carpenter had maintained its Nummuline
affinities before he obtained specimens showing this particular structure.   Structures so delicate as these are indeed only preserved exceptionally in fossil specimens.

3. To the cánal system, Professor Möbius does more justice, and admits its great resemblance to the forms of this structure in modern *Foraminifera.* This indeed appears from his own figures, as will be seen from the fac-simile tracings reproduced here, figs. 15 and 16, which well show how wonderfully this structure has been preserved, and how nearly it resembles the similar parts of modern *Foraminifera.* He thinks, however, that these round and regularly branching forms are rather exceptional, which is a mistake; though it is true that the sections of the larger canals are often somewhat flattened, and that they become flat where they branch. They are also sometimes altered by the vicinity of veinlets or fractures, or by minute mineral segregations in the surrounding calcite, accidents to which all similar structures in fossils are liable. Another objection, not original with him, is derived from their unequal dimensions. It is true that they are very unequal in size, but there is some definiteness about this. They are larger in the thicker and earlier formed layers, smaller or even wanting in the thinner and more superficial. In some slices the thicker trunks only are preserved, the slender branches having been filled with dolomite or calcite. It is difficult, also, to obtain, in any slice or any surface, the whole of a group of canals.* Farther, as I have shown, the thick canals sometimes give off groups of very minute tubes from their sides, so that the coarser and finer canals appear intermixed. These appearances are by no means at variance with what we know in other organic structures. Another objection is taken to the direction of the canals, as not being transverse to the laminæ but oblique. This, however, may be dismissed, since Möbius has of course to admit that it is not unusual in modern *Foraminifera.* It may be added that some of the appearances which puzzled Möbius, and which are represented in his figures, evidently arise from fractures displacing parts of groups of canals, and from the apparently sudden truncation of these at points where the serpentine filling gives place to calcite. It would also have been well if he had studied the canal system of those *Stromatoporæ* which have a secondary or supplemental skeleton, as *Cænostroma* and *Caunopora.*

4. A fatal defect in the mode of treatment pursued by Möbius is that he regards each of the structures separately, and does not sufficiently consider their cumulative force when taken together. In this aspect, the case of *Eozoon* may be presented thus : (1.) It occurs in certain layers of widely distributed limestones, evidently of aqueous origin, and on other grounds presumably organic. (2.)

---

* I have succeeded best in this by etching the surface of broken specimens.

Its general form, lamination and chambers, resemble those of the Silurian *Stromatopora* and its allies, and of such modern sessile foraminifera as *Carpenteria* and *Polytrema*. (3.) It shows under the microscope a tubulated proper wall similar to that of the Nummulites, though of even finer texture. (4.) It shows also in the thicker layers a secondary or supplemental skeleton with canals. (5.) These forms appear more or less perfectly in specimens mineralized with very different substances. (6.) The structures of *Eozoon* are of such generalized character as might be expected in a very early Protozoan. (7.) It has been found in various parts of the world under very similar forms, and in beds approximately of the same geological horizon. (8.) It may be added, though perhaps not as an argument, that the discovery of *Eozoon* affords a rational mode of explaining the immense development of limestones in the Laurentian age; and on the other hand, that the various attempts which have been made to account for the structures of *Eozoon* on other hypotheses than that of organic origin have not been satisfactory to chemists or mineralogists, as Dr. Hunt has very well shown.

Professor Möbius, in summing up the evidence, hints that Dr. Carpenter and myself have leaned to a subjective treatment of *Eozoon*, representing its structure in a somewhat idealized manner. In answer to this it is necessary only to say that we have given photographs, nature-prints and camera tracings of specimens actually in our possession. We have not thought it desirable to figure the most imperfect or badly preserved specimens, though we have taken pains to explain the nature and causes of such defects. Of course, when attempts at restoration have been made, these must be taken as to some extent conjectural; but so far as these have been attempted they have consisted merely in the effort to eliminate the accidental conditions of fossilized bodies, and to present the organism in its original perfection. Such restorations are not to be taken as evidence, but only as illustrations to enable the facts to be more easily understood. It is to be observed, however, that in the study of such fossils as *Eozoon*, the observer must expect that only a small proportion of his specimens will show the structures with any approach to perfection, and that comparison of many specimens prepared in different ways may be necessary in order to understand any particular feature. A single figure or a short description may thus represent the results of days spent in the field in collecting, of careful examination and selection of the specimens, of the cutting of many slices in different directions, and of much study of these with different powers and modes of illumination. My own collection contains hundreds of preparations of *Eozoon*, each of which represents perhaps hours of labor and study, and each of which throws some light more or less important on some feature of structure. The results of labor of this kind are unfortunately very liable to be regarded as subjective rather than objective by those who arrive at conclusions in easier ways.

Taken with the above cautions and explanations, the memoir of Professor Möbius may be regarded as an interesting and useful illustration of the structures of Eozoon, though from a point of view somewhat too limited to be wholly satisfactory.

The editor of the *American Journal of Science* gave Prof. Möbius an opportunity to reply, but stated that he had pledged himself that no rejoinder would be permitted—a somewhat unfair decision where the case was one of unprovoked attack, and this based on material published by Dr. Carpenter and myself. I protested against this, but in vain, and accordingly published my rejoinder in Canada, adding to it some remarks on other papers of similar nature. The following are extracts:—

Möbius has thought proper to take advantage of the security guaranteed to him by the editor of the *American Journal*, to reply to my courteous and somewhat forbearing criticism, in a manner which relieves me from any obligation to be reticent as to his errors and omissions. I shall, however, confine myself to those points in his rejoinder which seem most important in the interest of scientific truth.

1. With reference to the geological and mineral relations of *Eozoon*, I cannot acquit Möbius of a certain amount of inexcusable ignorance. More especially, he treats the structures as if they consisted merely of serpentine and calcite, and neglects to consider those specimens which, if more rare, are not less important, in which the fossil has been mineralised by Loganite, Pyroxene and Dolomite. If he had not specimens of these, he should have procured them before publishing on the subject. He neglects also to consider the broken fragments of *Eozoon* scattered through the limestones, and the multitudes of *Archæospherinæ* lying in the layers of deposit. Nor can I find that he has any clear idea how the structures of *Eozoon* could have been produced otherwise than by living organisms. Still farther, he makes requirements as to the state of preservation of the proper wall and canal system which would be unfair even in the case of Tertiary or Cretaceous *Foraminifera* injected with Glauconite, how much more in the case of a

very ancient fossil contained in rocks which have been subjected to great mechanical and chemical alteration.

2. In his reply he reiterates the statement that *Eozoon* is so different from existing *Foraminifera*, that, if this is a fossil, we must divide all organic bodies into " 1. Organic bodies with protoplasmic nature (all plants and animals); and 2. Organic bodies of Eozoonic nature (*Eozoon*, Dawson)." Without referring to the somewhat offensive way in which this is stated, I need only say that Dr. Carpenter has well replied that the structures of *Eozoon* are in no respect more different from those of modern *Foraminifera* than those of many other old fossils are from their modern representatives. All palæontologists know, for example, that while we cannot doubt that *Receptaculites, Archæocyathus,* and *Stromatopora* are organic, and probably Protozoan, it has proved most difficult to correlate their structures with those of modern animals.

3. I took occasion to mention certain errors of Prof. Möbius, due to his limited information on the subject of which he treats. He admits two of these, which were particularly pointed out, but taunts me with not producing others. This, however, would not have been difficult had I been disposed to enter in detail into a task so ungracious. Another example may be taken from his Plate XXXV, in which he represents together, and obviously for comparison, portions of the pores or tubuli of the modern *Polytrema,* and an imperfect fragment of the proper wall of *Eozoon,* and this more especially, as appears in the text, to show the comparative fineness of the latter. But the specimen of *Eozoon* is magnified only 75 diameters, while that of *Polytrema* is magnified 200 diameters, or in the proportion of 5625 to 40,000. Again, he has affirmed and repeats in his reply that the casts of the canal systems of *Eozoon* do not present cylindrical forms, but are "*flat* and *irregular* branched stalk-like bodies." If they appeared so to him, he must have possessed most exceptional specimens. Some canals, especially the larger, no doubt have flattened forms, particularly at their points of bifurcation; but this is comparatively rare, more especially in the vastly numerous minute canals, which are more frequently filled with dolomite than with serpentine. I have indeed been able to detect only a few out of very numerous specimens in which the majority of the casts of canals are not approximately round in cross section, even in the case of the larger canals. It is a question also if some flattening may not be due to pressure; and there are flat stolon-like tubes which can scarcely be called canals.*

---

* The forms of the canals are perhaps best seen in decalcified specimens; but Mr. Weston, who has done so much toward this investigation, has managed to cut slices so accurately at right angles to the general course of groups of canals, as to show their round cross sections with great distinctness. (See Fig. 6 *supra*.)

It occurs to me here to remark that Möbius seems to have over-looked the extremely fine canals injected with dolomite that fill the upper and thinner calcite walls of the better preserved specimens, and which in the thinner walls are nearly as fine as the tubuli of the proper wall, into which in many cases they almost insensibly pass where these last are themselves filled with dolomite. Possibly these structures have not been present in his specimens, or may have been destroyed or rendered invisible by his methods of preparation, and if so this would account for some of his conclusions. These fine canals are best seen in well-preserved serpentinous specimens free from chrysotile veins, and etched with very dilute nitric acid. They have scarcely been done justice to in any of the published figures either of Dr. Carpenter or myself, and do not appear in those of Prof. Möbius.

4. In reply to my objection that he has confounded the proper wall of *Eozoon* with veins of chrysotile, and that both are represented in his figures, he challenges me to point out which of the latter are chrysotile and which proper wall. Of course doing so will be of little importance to the argument, but I may indicate his figures 18, 43, 44 and 48 as in my opinion taken from portions of proper wall, and figure 45 seems to show the proper wall along with chrysotile. I may farther now point out to him that even Profs. King and Rowney in their recent paper admit that the proper wall is not continuous chrysotile, but consists of "aciculæ separated by calcareous interpolations," though they try to account for this structure by complicated changes supposed to have occurred in veins of chrysotile subsequently to their deposition.

In truth, the chrysotile veins cross all the structures of *Eozoon*, and those specimens are best preserved which have suffered least from this subsequent infiltration of chrysotile into cracks formed apparently by mechanical means. This has been amply shewn in figures which I have already published, but I have now still more characteristic specimens which may be seen in the Peter Redpath Museum.

5. Prof. Möbius sneers at my statement that when the proper wall of *Eozoon* is merely calcareous and not infiltrated, its structures are invisible, and that in many cases it has become opaque, while in thick slices its structure is always indistinct; but he should know that this is the case with all fine organic tubuli or pores in fossils penetrated with mineral matter, and eminently so with fossil Nummulites, as the researches of Carpenter have long ago demonstrated, and as any one possessing slices of these fossils can see for himself. I may add that in some decalcified specimens in my possession, where the proper wall has been wholly of calcite,

it is indicated merely by an *empty band* intervening between the serpentine cast and the supplemental skeleton filled with casts of canals.

6. Lastly, he seems to think that no offence should be taken at his insinuation that the figures printed by Dr. Carpenter and myself are idealized or untruthful representations, and he repeats the accusation in the following terms: "The individual peculiarities of diagrams should not exceed the limits of the known variability of the real specimens, but in the *Eozoon* diagrams of Carpenter and Dawson these limits are exceeded." There could not, I think, be a more plain charge of wilful falsification, and this is made by a naturalist who discusses *Eozoon* without having taken the pains to study it *in situ*, or to avail himself of the large collections of specimens which exist in England and in Canada. I can only reply that while I have been unable to figure all the peculiarities of the canal systems of this complicated and often badly preserved fossil, I have endeavoured to select the most characteristic specimens,—and that my representations are principally nature-prints, photographs and camera tracings, some of the latter by artists in no way interested in *Eozoon*. Dr. Carpenter's representations appear to me to be equally truthful. Neither of us have taken the trouble to represent badly preserved or imperfect specimens, any more than we should do so in the case of any other fossil, when better examples were procurable.

In connection with this, Möbius seems to think that in my criticism I should have gone into all the details into which he enters. This was unnecessary, except to expose his principal errors or misstatements. It could not have been done without publishing a treatise as long and as expensively illustrated as his own,—and this I should prefer to do in some other form than as a mere reply to him, and with reference to much larger and more varied collections than those at his command.

He is good enough to add that if I will send him more and better specimens, he will willingly "forgive" me for "disappointing" him and other naturalists. I must say that I cannot purchase forgiveness on such terms, but if he will take the trouble to visit Canada and inspect my collections, he shall have every opportunity to do so.

I think it is only due to the interests of palæontological science to add here, that I attach more blame to the editors of the German publication "Palæontographica," in which his memoir appears, than to Prof. Mobius himself. We have been in the habit of regarding this publication as one in which the matured results of original observers and discoverers are given, and when it devotes forty costly plates to the labours of a naturalist who is not of this character, in so far as *Eozoon* is concerned, and who has not even studied the principal collections on which other naturalists equally competent have based their conclusions, they incur a responsibility much more grave than if they were merely the conductors of a popular scientific journal, open to cursory discussions of controverted points. They cannot relieve themselves from this responsibility till they shall have published a really exhaustive description of *Eozoon* by some of the original workers on the subject. This is the more necessary, since if *Eozoon* is really a fossil, its discovery is one of the most important in modern palæontology, and since its claims cannot be settled except by the most full investigation and discussion.*

Still later than the reply of Möbius, are two additional papers of still more remarkable character. For, while Möbius is content to take up a purely negative position, these undertake to account for the structures of *Eozoon* by other causes than that of animal growth, and by causes altogether inconsistent with one another. The first of these is an abstract of a memoir " On the origin of the mineral, structural and chemical characters of Ophites and related rocks," presented to the Royal Society of London by Professors King and Rowney. The second is a quarto pamphlet of 96 pages with 30 plates, by Dr. Otto Hahn, entitled " Die Urzelle," the " Primordial cell."

The first of these papers contains little that is new, being a re-habilitation of that hypothesis of " Methylosis," or chemical transmutation, which the authors have already fully explained in the Transactions of the Irish Academy and elsewhere. Its bearing on *Eozoon* is simply this :—that if any one acquainted with geological and chemical possibilities can be induced to believe that the Laurentian limestones of Canada are " Methylosed products," which originally " existed as gneisses, hornblende schists, and other mineralised silacid metamorphics," he may be induced also to believe that *Eozoon* is a product of merely mineral metamorphism.

When we consider that these great limestones have been so fully traced and mapped by Sir William Logan and his successors on the Geological Survey ; that some of them are several hundreds of feet in thickness and traceable for great distances, that they are quite comformable with the containing beds, and themselves exhibit alternating layers of limestone and dolomite, with layers characterized by the presence of graphite, serpentine, and other

---

* This complaint was published several years ago, and sent to the editors of the publication in question, without eliciting any reply or redress. They thus lay themselves open to the charge of wilful falsification.

minerals, and subordinate thin bands of gneiss and pyroxene rock, the idea that they can be products of a sort of pseudomorphism of gneisses and similar rocks, becomes stupendously absurd, and can only be accounted for by want of acquaintance with the facts on the part of the authors.

To explain the structures of *Eozoon*, however, even this is not altogether sufficient, but we must suppose a peculiar and complex arrangement of laminæ, canals, and microscopic tubuli or fibres simulating them, to be produced in some parts of the limestones and not in others ; and this by the agency of several different kinds of minerals.

In other words we have to suppose a conversion on a gigantic scale of gneiss into dolomite, limestone, graphite, serpentine, and other minerals, consisting for the most part even of different elements, and this at the same time or by still more mysterious subsequent changes, producing imitations of the most delicate organic forms. The mere statement of this hypothesis is, I think, sufficient to show that it cannot be accepted either by chemists or palæontologists, and it only serves to illustrate the difficulties which *Eozoon* presents to those who will not accept the theory of its organic origin.

Dr. Otto Hahn regards the matter from an entirely different point of view. He has himself visited Canada, has collected specimens of *Eozoon*, and now proposes to effect an entire revolution in our ideas of the palæontology of the Eozoic rocks.

In a former paper he had maintained that *Eozoon* is altogether of mineral origin, that its serpentine is hydrated olivine, and the canal system merely cracks in calcite injected by the expansion of this mineral. This hypothesis he now finds untenable, and he regards *Eozoon* as a vegetable production, or rather as a series of such productions. He regards the laminæ as petrified fronds of a sea-weed, and the canal systems as finer algæ of several genera and species. Not content with this, he describes as plants other forms found in granite, gneiss, basalt, and even meteoric iron, and others found included in the substance of crystals of Aragonite, Corundum and Beryl. All these are supposed to be algæ of new species, and science is enriched by great numbers of generic and specific names to designate them, while they are illustrated by thirty plates representing the quaint and grotesque forms of these objects, many of which are obviously such as we have been in the habit of regarding as mere dendritic crystallisation, cavities, or impurities included in crystals.

It seems scarcely necessary to criticise such views, as it is probable that very few naturalists will be disposed to accept the

supposed plants described by Dr. Hahn as veritable species. It
may be observed, however, that in regarding the thick plates of
serpentine, interrupted, attached to each other at intervals,
penetrated by pillars of calcite, and becoming acervuline upward,
as fossil algæ, he disregards all vegetable analogies; while in
supposing that the calcite is a filling, and that the delicate fillings
of canals contained in it are fine thread-like algæ, he equally
asserts what is improbable. Farther, no vegetable structure or
remains of carbonaceous matter have been discovered in the
serpentine. Had he discovered these supposed vegetable forms in
the graphite of the Laurentian, this would have been far more
credible.

Hahn's paper, however, suggests one or two points of interest
respecting *Eozoon*, which have perhaps not been sufficiently insisted
on. One of these is the occurrence of rounded " chamberlets " in
the calcareous walls. These are his " germ-cells," and they sometimes present the curious character that they are hollow vesicles of
serpentine filled with calcite, and when these have been cut across
in making a section, and the calcite has been dissolved out with an
acid, they present very singular appearances. They may in some
cases have been germs of *Eozoon*, or smaller foraminifera of the
type of *Archæospherinæ*, overgrown by the calcareous walls. It is
farther to be observed, as I have also elsewhere remarked, that the
serpentine, filling the larger spaces between the calcareous laminæ
sometimes shows indications of deposit as a lining of the cells, and
in some specimens this lining has not filled the original space, but
has left a drusy cavity afterwards filled with calcite.

Again, in parts of the canal system, especially when filled with
dolomite, there occur little disc-like bodies or trumpet-shaped
terminations of canals. These, I fancy, are the calyx-like objects
figured by Hahn. Their precise significance is not known, further
than that they may represent the expanded ends of canals. Another
appearance deserving of notice is the occurrence of portions of
specimens of *Eozoon* in which little or no serpentine occupies the
chambers. In this case the laminæ have either been pressed close
together, or the chambers have been filled with calcite not distinguishable from the walls, in which, however, the casts of groups,
of canals often occur, and might then be more readily mistaken for
algæ than when they occur between laminæ of serpentine.*

It may be said, in connection with the attacks in question, that
if *Eozoon* is an object of which so many and strange explanations

---

* I have not thought it necessary to refer to still more recent papers of other
German writers, which add nothing material to the discussion of the question.

can be given, it is probable that no certainty whatever can be attained as to its real nature. On the other hand it is fair to argue that, if the opponents of its animal nature are driven to misrepresentation and to wild and incoherent theories, there is the more reason to repose confidence in the sober view of its origin, consistent with its geological relations and microscopic characters, which has commended itself to Carpenter, Gümbel, Rupert Jones, Sterry Hunt, and a host of other competent naturalists and geologists. For my own part, the arguments adduced by opponents, and the re-examination of specimens which they have suggested, have served to make my original opinion as to its nature seem better supported and more probable; though of course I would be far from being dogmatic on such a subject, or claiming any stronger conclusion than that of a reasonable probability, which may be increased as new facts develop themselves, but cannot amount to absolute certainty until the discovery of Laurentian rocks in an unaltered state shall enable us to compare their fossils more easily with those of later formations.

In point of fact, the evidence for the organic nature of a fossil such as that in question, is necessarily cumulative, and depends on its mode of occurrence and state of mineralisation, as well as on its general form and microscopic structure; and it is perhaps hopeless to expect that any considerable number of naturalists will be induced to undertake the investigations necessary to form an independent opinion on the subject. It may be hoped, however, that they will fairly weigh the evidence presented, and will also take into consideration the difficulty of accounting for such forms and structures except on the hypothesis of an organic origin.

VII. PALÆOZOIC FOSSILS ASSOCIATED WITH SERPENTINE AND OTHER HYDROUS SILICATES AS ILLUSTRATIVE OF EOZOON.*

Fossils having their cavities and pores infiltrated with hydrous silicates are much more abundant in Palæozoic limestones than is usually imagined. In some instances, serpentine itself is found to have been concerned in such infiltration; while in other cases, the infiltrating hydrous

---

* This is in substance from a paper published in the Journal of the Geological Society, 1879. The microscopic cabinet of the Peter Redpath Museum possesses a large collection of these infiltrated specimens.

silicates are found to approach to jollyte, fahlnnite, and other minerals whic h have usually been regarded as products of decomposition or metamorphism, but which, as Dr. Sterry Hunt has justly remarked, cannot reasonably be referred to such an origin when they are found filling the pores of Crinoids and other fossils in strictly aqueous deposits. In this case they must surel y be the results of original deposition in the manner of glauconite ; and, as we shall find, they sometimes appear to be strictly the representatives of that mineral, which occurs under similar conditions in other parts of the same formation.

(1.) *Serpentine of Lake Chebogamong.*

Mr. Richardson, of the Geological Survey, has observed, north of the Laurentian axis, on the Saguenay River, certain rocks which appear to be similar in mineral character to the Quebec group of Sir William Logan, and occupy a geological position intermediate between the Laurentian and the Trenton formation. Among these, he describes a band of serpentine associated with limestone at Lake Chebogamong, which lies about 200 miles to the N.E. of Lake St. John, in a little-explored region. Among the few specimens which Mr. Richardson was able to bring back with him was one of extreme interest—a specimen apparently from the junction of the limestone and serpentine, and containing a portion of a tabulate coral, of which some of the cells are filled with a mixture of serpentine and calcite, and some with calcite. The serpentine seems to have been weathered; it has a granular, uneven appearance, and under the microscope shows patches with fibrous structure like chrysotile. There are also whitish serpentine veins, fringed with chrysotile or a mineral resembling it under the microscope. The cell-walls of the coral are perfectly black and opaque, and probably carbonaceous. The coral found thus mineralized was examined by Mr. Billings, who had no doubt of its nature, though uncertain as to its generic affinities. After careful study of it, I am disposed to refer it to the genus

5

Astrocerium of Hall, and it is not distinguishable in structure from *A. pyriforme* of that author, a species very common in the Upper Silurian limestones of the region in which the specimen occurs. The genus *Astrocerium* is specially characteristic of the Niagara formation; and though Edwards doubts its distinctness from *Favosites*, I think there are constant points of difference, especially in the microscopic characters of the cell-walls, which entitle it to be separated from that genus. In such specimens of *Astrocerium* as are well preserved, the walls of the hexagonal cells seem to have been of corneous texture, with minute corneous spicules instead of radiating septa. They have pores of communication, and there are also occasional larger pores or tubes in the angles of the cells. The tabulæ are very thin and apparently purely calcareous. This accounts for the singular fact, mentioned by Hall, that the cell-walls are sometimes entirely removed, leaving the tabulæ in concentric floors like those of *Stromatopora*. I think it likely that the typical species of *Astrocerium* may have been inhabited by Hydroids, and may have been quite remote from *Favosites* in their affinities.

The formation in which the serpentine and limestone of Lake Chebogamong occur, is described as consisting of chloritic slates in some places with hornblende crystals, dolomites, and hard jaspery, argillaceous rocks. Upon these rest conglomerates and breccias with Laurentian fragments, and also fragments of the rocks before mentioned, and on these lie the limestone and serpentine. The serpentine has been analysed by Dr. Hunt, who finds it to contain chromium and nickel, and in this respect to be similar to that of the Quebec group, and not to that of the Laurentian.* The fossil would give evidence of a much later date than that usually attributed to rocks of the character above stated; but it is quite possible that there may be two series of different ages in the region, the lower being Lower Silurian or perhaps older, and the upper of Upper Silurian age. If the serpentine belongs to the newer formation, its association with a coral of the genus *Astrocerium* would of course be quite natural. If it belongs to the older formation, and the overlying limestone to the newer, the serpentine in the latter may be a *remanié* silicate derived from the older rocks and mixed with the limestone at their junction.

(2.) *Serpentine of Melbourne.*

The serpentines of this place belong to a great series of more or less altered rocks extending through the province of Quebec, and referred by Sir William Logan, on strati-graphical grounds, to his Quebec group, equivalent to the Arenig or Skiddaw series of England.[†] In ascending order, these rocks at Melbourne present first a thick series of highly plumbaginous schists or shales, with thin bands of limestone holding fragments of Lower Silurian corals and crinoids. These pass upwards into a thick series of slaty rocks characterized by the prevalence of a shining crystal-line hydro-mica, and known as nacreous or hydro-mica slates. They are associated with quartzose bands, and also with lenticular layers of crinoidal limestone. Parallel with these beds and, according to Logan's observations, over-lying them, though later observers regard the junction as faulted, is the series containing the serpentine, which is associated with layers of limestone and nacreous slate, and also with brecciated and arenaceous beds, probably originally tufaceous, with beds of anorthite, steatite, and dolomite, and also with red slates, the whole forming a miscellaneous and irregular group, evidently resulting from the contemporaneous action of igneous and aqueous agencies, and affording few traces of fossils. The serpentines, which occur in thick and irregular beds, are different in colour and microscopic texture from those of the Laurentian system, and also present some chemical differences, more especially in the presence of oxides of nickel, chromium and

---

* Report of Geological Survey of Canada, 1870-71.

† Selwyn and Hunt, however, hold that these rocks are in part Huronian.

cobalt, and of a large percentage of iron and a smaller proportion of water.*

These serpentines are undoubtedly bedded rocks and not eruptive ; but they may have originated from the alteration of volcanic materials.† They appear, shortly after their original deposition, to have been broken up, so as in many places to present a brecciated appearance, the interstices of the fragments being filled with limestone and dolomite, which themselves are largely mixed with the flocculent serpentinous matter, and traversed by serpentinous veins sometimes compact and sometimes fibrous. Besides the very impure limestone thus occurring in the serpentinous breccia, there are also true layers or beds of limestone and dolomite included in or near to the great serpentine band. No well-preserved fossils have been found either in these beds or in the brecciated serpentine; but on treating the surfaces of slabs with an acid or making thin slices, fragments of organic bodies are developed which well illustrate the manner in which serpentine, whatever its origin, may be connected with the mineralization of such fragments.

It is to be observed here that the irregular bedding of the serpentine, and the apparent passage on the line of strike into dolomite and red slate, might accord either with a purely aqueous and oceanic mode of deposition like that of glauconite, or with deposition as beds of volcanic sediments, afterwards altered and partly re-deposited by water. The association with ash rocks and agglomerates would however, tend rather to the latter view, as would also the chemical characters of the serpentine already referred to;

* Under the microscope, the Laurentian serpentines are usually homogeneous and uncrystalline, with the structure of netting veinlets which I have elsewhere called septariform. The Melbourne serpentines usually present a confused mass of acicular crystals or a fibrous structure, and, where structureless, polarize more vividly than those of the Laurentian.

† Sandborger (Essay on Metallic Veins) quotes many German chemists to the effect that "olivine rock and the serpentine formed from it always contain copper, nickel, and cobalt." This origin might thus apply to the serpentines in the Quebec group in Canada, but not to those of the Laurentian, as I have already urged on other grounds in my reply to Hahn, in the "Annals and Magazine of Natural History," 1876, vol. xviii. pp. 32, 33.

but the association with fossils mentioned below tends to show that at least a part of the mineral is an ordinary aqueous deposit. It is also to be observed, with reference to the superposition of serpentine on fossiliferous Lower Silurian rocks, that a similar relation is affirmed by Murray to occur in Newfoundland, where massive serpentines overlie unaltered fossiliferous rocks of the Quebec group.

No fossils have been found in the compact serpentine, but only in the limestone paste of the brecciated masses and in the limestone bands interstratified. The limestone of the breccia contains not only angular fragments of serpentine but disseminated serpentine and small veins of the same mineral. Its fossils are limited to small tubular bodies, crinoidal joints, and fragments, apparently of *Stenopora*, very imperfectly preserved. The tubular bodies may be portions of *Hyolithes* or *Theca*. Their interior is usually filled with dolomite; their walls are in the state of calcite; and they are incrusted with an outer ring of serpentine. In some instances the calcareous organic fragments are seen to be filled in the interior with serpentine. The crinoidal fragments are in a similar condition, the serpentine having apparently surrounded them in a concretionary manner after the cavaties had been filled with dolomite. Fragments of calcite, dolomite, or older serpentine included in the limestone, and of no determinate form, are enclosed in the new or *remanié* serpentine in like manner, and in some cases this newer or coating serpentine was observed to have a fibrous structure. The serpentine thus coating and filling fossils and fragments is of a lighter colour than the serpentine of the fragments themselves, and in this respect resembles that of the small veins traversing the limestone. Such traces of fossils as exist in the layers of limestone are similar to those in the breccia, but not, so far as observed, coated with serpentine.

It would thus appear that, contemporaneously with the original deposition of the serpentine, thin bands of limestone were laid down, with a few fragments of crinoids, corals, and shells; that subsequently, but perhaps within the same geological period, and while the deposition of serpentine was still proceeding, portions of the surface of the serpentine were broken up and imbedded in limestone; that the fissures of this limestone were penetrated with serpentine veins, and its few fossils coated with that mineral, which also forms flocculent laminæ in the limestone.

The mode of deposition of this Palæozoic serpentine is thus considerably different from that of the Laurentian, which forms layers intimately interstratified with great limestones, and also nodules, concretionary grains, and fillings of fossils in these limestones. This difference in mode of occurrence is, no doubt, connected with the difference in composition of the two varieties of the mineral already noticed. In both cases, however, the serpentine has been so deposited that it could take part in the mineralization of marine organic remains.

The condition of the fragments of Silurian fossils in the limestones associated with the nacreous or hydro-mica slates is of interest in connexion with this subject. The shining laminated mineral associated with these fossils has been regarded from its chemical composition as a hydro-mica. Under the microscope, however, it shows a want of homogeneity which suggests the presence of two or more silicates, or the association of crystals of hydrous mica with minute grains of silicious matter of some other kind. Though now highly crystalline, it must originally have been a fine sediment, since it fills the finest cells of *Stenopora* and *Ptilodictya*. Nor can its present state have been produced by any extreme metamorphism, as the undistorted state of these fossils amply testifies. Further it is interesting to observe that though the hydrous silicate is little magnesian, the fossils themselves are not infrequently converted into dolomite. In these fossiliferous beds there are also tabular crystals, apparently of anhydrous mica, little groups of crystals of tremolite, cavities filled with quartz, and crystalline grains of a mineral having the microscopical characters of olivine; and these have been developed or included in the mass without injury to the structures of the most delicate corals.

Similar appearances are presented by limestones from other parts of the Quebec group, of which a great series of slices has been prepared by Mr. Weston under the direction of the late Sir W. E. Logan, who, in his later researches in this group of rocks, gave much attention to the microscopic fossils in the more altered beds, as a means of determining their ages. Besides large series from Melbourne and its neighbourhood, I have examined slices from Stanford, Farnham, Cleveland, Bedford, Orford, Arthabaska, Point Levi, Rivière du Loup, and other places, in most of which Lower Silurian fossils occur associated with hydrous silicates.

The fossils above referred to occur in rocks undoubtedly of Lower Silurian age, and regarded as altered or metamorphosed members of the Quebec group. In the unaltered representatives of these rocks at Point Levi and the Island of Orleans there occur considerable quantities of a true glauconite, which has been analyzed by Dr. Hunt, and which is without doubt an original deposit in the sandy and argillaceous beds in which it occurs, which in many cases are precisely similar to Cretaceous greensands. Dr. Hunt's analysis shows that this glauconite contains alumina, iron, potash, and magnesia, and thus approaches to the Laurentian loganite. In the forms of its little concretions it resembles the serpentine grains in the Laurentian limestones; and like modern glauconite it has moulded itself in organic forms. Some of these are spiral or multilobate, as if casts of minute univalve shells or of spiral and textularine Foraminifera.* Others are annular or are arcs of circles, and some present a delicate fibrous or tubulated appearance, as if they had moulded themselves on porous shells or very minutely-celled corals, spicules of sponges, &c. Shreds of corneous Polyparies, perhaps of Graptolites, abound in the matrix, but are not connected with the glauconite grains. Unfortunately there are no *Stromatoporæ* in these beds,

* Ehrenberg has found casts of rotaline and textularine Foraminifera in Lower Silurian beds in Russia; and such forms occur in Upper Silurian limestones in Nova Scotia.

otherwise we might have an almost precise recurrence of
of the relations of serpentine with *Eozoon* in the Laurentian.*

Another appearance which may be mentioned in this
connexion occurs in certain beds of Utica Slate in the
vicinity of the trappean mass of Montarville, and converted
into a hard sonorous rock. In one of these are stems of
crinoids which have retained their external form, while the
calcareous material has been entirely removed and replaced
by a soft green crystalline mineral whose physical and microscopical characters are those of chlorite, and which in
any case may be regarded as one of those hydrous silicates
sometimes termed " viridite."

(3.) *Limestone of Pole Hill, N.B., and of Llangwyllog in
Anglesey.*

In a paper in the Transactions of the Royal Irish Academy,
and subsequently in " Life's Dawn on the Earth," I noticed
a remarkable limestone discovered by Mr. C. Robb, of the
Geological Survey, at Pole Hill in New Brunswick, and
believed to be of Upper Silurian age. It is composed of
fragments of crinoids and shells, the cavities of which are
finely injected with a hydrous silicate of alumina, iron, and
magnesia, the composition of which, according to Dr. Hunt,
approaches to that of the jollyte of Von Kobell, and also to
that of a hydrous silicate described by Hoffmann as filling
the cavities of specimens of *Eozoon* found in Bohemia. It
has also some resemblance to the loganite which mineralizes
the *Eozoon* of Burgess, in Canada. At the same time I
mentioned a specimen of limestone of similar character
which I had found in the McGill College collection, and
which I supposed to be from Wales. It is labelled " Llan-
golloc," and belonged to the collection of the late Dr. Holmes,
of Montreal. Sir A. Ramsay, to whom I have applied
for information as to the locality, kindly informs me that
the name is probably " Llangwyllog," that the place so
named is in Anglesey, and that limestone of Lower Silurian
or Cambrian age occurs in its vicinity.

* Report of Geological Survey of Canada, 1866.

A portion of this silicate was submitted to Dr. Sterry Hunt, from whose analysis it appears to be of similar character with that of Pole Hill, and like it injects in the most beautiful manner the pores and cavities of crinoids, shells, and corals.* The limestone containing this silicate is of subcrystalline texture, with occasional bright cleavage-faces which belong to crinoidal fragments. Its colour, owing to the included silicate, is dull olive, and it shows occasional small deep green and reddish specks. Its aspect is so waxy, that at a little distance it might be mistaken for an impure serpentine.†

When examined with the microscope, the flocculent olive-green silicate is seen to penetrate the mass exactly in the

---

* As the analyses of these specimens by Dr. Hunt are of much interest, they are given in the following table :—

	Pole Hill, New Brunswick	Llangwyllog, Wales.
Silica.......................	38-93	35-32
Alumina.....................	28-88	22-66
Protoxide of iron..........	18-86	21-42
Magnesia...................	4-25	6-98
Potash.....................	1-69	1-49
Soda.......................	0-48	0-67
Water......................	6-91	11-46
	1000-00	100-00

In the Llangwyllog specimen the silicate amounted to three per cent. of the whole, the remainder being carbonate of lime with a very little siliceous sand and fine clay. In the Pole-Hill specimen the silicate amounted to about five per cent., the remainder being limestone with a few quartz grains.

It will be seen that these two silicates, evidently deposited from solution in such a manner as to fill the finest organic pores, are remarkably similar in composition ; and the fact that they closely resemble Hoffmann's mineral found in Bohemian *Eozoon*, and also the loganite filling the Burgess *Eozoon* (Quart. Journ. Geol. Soc. vol. xxi. 1865), gives them additional interest.

† In the Map illustrating Blake's paper on the "Monian" of Anglesey, Llanwyllog, is placed on Ordovician rocks near their junction with older Chloritic Schists. Jour. Geol. Socy., Aug., 1888.

manner of the serpentine in ophiolite, and it has a polariscope appearance approaching to that of serpentine; while greenish by reflected light, it appears reddish when seen in thin slices with transmitted light. It penetrates the finest pores of crinoids, and at the same time fills the cavities of shells and the cells of corals. The larger fillings of this kind give the deep green spots above mentioned, while the red spots are apparently caused by the partial oxidation of the iron of the mineral. In one shell, apparently a small *Orthoceras* or *Theca*, the dark green filling has cracked in the manner of *Septaria*, and the fissures have been filled with carbonate of lime. In some places the mineral has penetrated the pores of shells of Brachiopods or crusts of Trilobites, producing a tubulated appearance not unlike the proper wall of *Eozoon*.

From the characters of the fragments, I should imagine that this limestone is Lower Silurian rather than Cambrian. It affords an excellent instance of the occurrence of hydrous silicates infiltrating organic fragments, and it deserves the attention of collectors having access to the locality. A curious point of coincidence of the limestone with some of those in the Lower Silurian of Canada is the occurrence of a few bright green specks, probably of apatite or vivianite, giving on a small scale that association of phosphates with hydrous silicates which we find on the great scale in the Laurentian.

I have in a previous part of this paper referred to the remarkable specimens from Maxville which show in the Carboniferous age infiltrations with a hydrous silicate comparable to these in the older and later formations. We thus find that such infiltration is a common fact all the way from the Laurentian to the modern series, and the circumstance that the silicates employed may in different cases be those of magnesia, iron, alumina, or potash, or mixtures of these, does not seriously affect the significance of this fact.

VIII. THE PHOSPHATES AND GRAPHITE OF THE LAURENTIAN AND CAMBRIAN ROCKS OF CANADA.

Apatite of the Laurentian, etc.

The extent and distribution of the deposits of apatite contained in the Laurentian of Canada and in the succeeding Palæozoic formations, have not escaped the notice of our Geological Survey, and have been referred to in some detail in reports of Mr. Vennor, Mr. Richardson, and others, as well as in the General Report prepared by Sir W. E. Logan in 1863. Some attention has also been given, more especially by Dr. Sterry Hunt, to the question of the probable origin of these deposits.* My own attention has been directed to the subject by its close connexion with the discussions concerning *Eozoon ;* and I have therefore embraced such opportunities as offered to visit the localities in which phosphates occur, to examine their relations and structure and to collect illustrative specimens.† I would now present some facts and conclusions respecting these minerals, more especially in their relation to the life of the Laurentian period.

In the Cambrian and Lower Silurian rocks of Canada phosphatic deposits occur in many localities, though apparently not of sufficient extent to compete successfully for commercial purposes with the rich Laurentian beds and veins of crystalline apatite.

The Acadian or Menevian group, as developed near St. John, New Brunswick, contains layers of calcareous sandstone blackened with phosphatic matter, which can be seen under the lens to consist entirely of shells of *Lingulæ*, often entire, and lying close together in the plane of the deposit, of which in some thin layers they appear to constitute the principal part.‡ Mr. Matthew informs me that these layers

---

* The Museum also contains large suites of specimens collected by Dr. Harringtons and others purchased with the collections of the late Mr. J. G. Miller.

† Geology of Canada, 1863; Chemical and Geological Essays, 1875.

‡ Bailey and Matthew, "Geology of New Brunswick," Geol. Survey Reports.

belong to the upper part of the formation, and that the layers crowded with *Lingulæ* are thin, none of them exceeding two inches in thickness; but he thinks that the dark colour of some of the associated sandstones and shales is due to comminuted *Lingulæ*.

In the Chazy formation, at Alumette Island, and also at Grenville, Hawkesbury, and Lochiel, dark-coloured phosphatic nodules abound. They hold fragments of *Lingulæ*, which also occur in the containing beds. They also contain grains of sand, and, when heated, emit an ammoniacal odour. They are regarded by Sir W. Logan and Dr. Hunt as coprolitic, and are said to consist of "a paste of comminuted fragments of *Lingulæ*, evidently the food of the animals from which the coprolites were derived." * It has also been suggested that these animals may have been some of the larger species of Trilobites. In the same formation, at some of the above places, phosphatic matter is seen to fill the moulds of shells of *Pleurotomaria* and *Holopea*.

In the Graptolite shales of the Quebec group, at Point Levis, similar nodules occur; and they are found at Rivière Ouelle, Kamouraska, and elsewhere on the Lower St. Lawrence, in limestones and limestone conglomerates of the same series. In these beds there are also small phosphatic tubes with thick walls, which have been compared to the supposed worm-tubes of the genus *Serpulites*.†

At Kamouraska, where I have studied these deposits, the ordinary phosphatic nodules are of a black colour, appearing brown with blue spots, when examined in thin slices with transmitted light. They are of rounded forms, having a glazed but somewhat pitted surface—and are very hard and compact, breaking with glistening surfaces. They occur in thin bands of compact or brecciated limestone, which are very sparingly fossiliferous, holding only a few shells of *Hyolithes* and certain *Scolithus*-like cylindrical markings. In some of these beds siliceous pebbles occur with the

* Geology of Canada, p. 125.
† Geology of Canada, p. 259; Richardson's Report, 1869.
§ Journal Geological Society, 1876.

nodules, rendering it possible that the latter may have been derived from the disintegration of older beds; but their forms show that they are not themselves pebbles. Phosphatic nodules also occur sparingly in the thick beds of limestone conglomerate which are characteristic of this formation; they are found both in the included fragments of limestone and in the paste. The conglomerates contain large slabs and boulders of limestone rich in Trilobites and *Hyolithes;* but in these I have not observed phosphatic nodules.

In some of the limestones the phosphatic bodies present a very different appearance, first noticed by Richardson at Rivière Ouelle, and of which I have found numerous examples at Kamouraska. A specimen now before me is a portion of a band of grey limestone, about four inches in thickness, and imbedded in dark red or purple shale. It is filled with irregular, black, thick-walled, cylindrical tubes, and fragments of such tubes, along with phosphatic nodules —the whole crushed together confusedly, and constituting half of the mass of the rock. The tubes are of various diameters, from a quarter of an inch downward; and the colour and texture of their walls are similar to those of the ordinary phosphatic nodules.

Under the microscope the nodules and the walls of the tubes show no organic structure or lamination, but appear to consist of a finely granular paste holding a few grains of sand, a few small fragments of shells without apparent structure, and some small spicular bodies or minute setæ. The general colour by transmitted light is brown; but irregular spots show a bright blue colour, due probably to the presence of phosphate of iron (vivianite). The enclosing limestone and the filling of the tubes present a coarser texture, and appear made up of fragments of limestone and broken shells, with some dark-coloured fibres, probably portions of Zoophytes. Scattered through the matrix there are also small fragments, invisible to the naked eye, of brown and blue phosphatic matter.

One of the nodules from Alumette gave to Dr. Hunt

36-38 of calcic phosphate; one from Hawkesbury 44-70; another from Rivière Ouelle 40-34: and a tube from the same place 67-53.* A specimen from Kamouraska, analyzed by Dr. Harrington, gave 55-65 per cent. One of the richest pieces of the linguliferous sandstone from St. John yielded to the same chemist 30-82 of calcic phosphate and 32-44 of insoluble siliceous sand, the remainder being chiefly carbonate of lime.

Various opinions may be entertained as to the origin of these phosphatic bodies; but the weight of evidence inclines to the view originally put forward by Dr. Hunt,† that the nodules are coprolitic; and I would extend this conclusion with some little modification to the tubes as well. The forms, both of the tubes and nodules, and the nature of the matrix, seem to exclude the idea that they are simply concretionary, though they may in some cases have been modified by concretionary action. There are in the same beds little piles of worm-castings of much smaller diameter than the tubes, and less phosphatic; and there are also *Scolithus*-like burrows penetrating some of the limestones, and lined with thin coatings of phosphatic matter similar to that of the tubes. Further, the association of similar nodules in the Chazy limestone with comminuted *Lingulæ*. as already stated, is a strongly confirmatory fact.

The tubes are of unusual form when regarded as coprolitic; but they may have been moulded on the sides of the burrows of marine worms; or these creatures may have constructed their tubes of this material, either consisting of their own excreta or of that of other animals lying on the sea-bottom. In any case, the food of the animals producing such excreta must have been very rich in solid phosphates, and these animals must have abounded on the sea-bottoms on which the remains have accumulated. It is also evident that such phosphatic dejections might either retain their original forms, or be aggregated into nodular masses, or shaped into tubes or burrows of Annelids, or, if accumulated in mass, might form more or less continuous beds.

* Geology of Canada, p. 461

The food of the animals producing such coprolites can scarcely have been vegetable; for though marine plants collect and contain phosphates, the quantity in these is very minute, and usually not more than required by the animals feeding on them.

We must therefore look to the animal kingdom for such highly phosphatic food. Here we find that a large proportion of the animals inhabiting the primordial seas employed calcic phosphate in the construction of their hard parts. Dr. Hunt has shown that the shells of *Lingula* and some of its allies are composed of calcic phosphate; and he has found the same to be the case with certain Pteropods, as *Conularia*, and with the supposed worm-tubes called *Serpulites*, which, however, are very different in structure from the tubes above referred to.

It has long been known that the crusts of modern Crustaceans contain a notable percentage of calcic phosphate; and Hicks and Hudleston have shown that this is the case also with the Cambrian Trilobites. Dr. Harrington has kindly verified this for me by analyzing a specimen of highly trilobitic limestone, probably Middle Cambrian, from St. Simon, in which the crusts of these animals are so well preserved that they show their minutely tubulated structure in great perfection under the microscope.* He finds the percentage of calcic phosphate due to these crusts to be 1.49 per cent. of the whole mass. It is to be observed, however, that the crusts of Trilobites must have consisted very largely of chitinous matter, which, in some cases, still exists in them in a carbonized state. A crust of the modern *Limulus*, or King Crab, which I had supposed might resemble in this respect that of the Trilobites, was analyzed also by Dr. Harrington. It belonged to a half-grown individual, measuring 5·25 inches across, and was found to contain only 1·845 per cent. of ashes, and of this only 1·51 per cent. of calcic phosphate. The crusts of some Trilobites may have contained as large a proportion of organic matter; but they would seem to have been richer in phosphates.

* This limestone is from a large boulder in the Quebec group conglomerate.

Next to *Lingulæ* and Trilobites, the most abundant fossils
in the formations containing the phosphatic nodules are the
shells of the genus *Hyolithes*, of which several species have
been described by Mr. Billings.*   Dr. Harrington has
ascertained that these shells also contain calcic phosphate
in considerable proportion.† The proportion of this substance
in a shell not quite freed from matrix was 2·09 per cent.
These shells have usually been regarded as Pteropods; but
I find that the Canadian primordial species show a structure
very different from that of this group. They are much
thicker than the shells of proper Pteropods; and the outer
layer of shell is perforated with round pores, which in one
species are arranged in vertical rows. The inner layer,
which is usually very thin, is imperforate. In one species
(I believe, the *H. americanus* of Billings) the perforations
resemble in size and appearance those in the shells of
Terebratulæ. In another species (*H. micans* probably) they
are very fine and close together, as in some shells of
tubicolous worms. I am therefore disposed to regard the
claim of these shells to the rank of Pteropods as very
doubtful. They may be tubicolous worms, or even some
peculiar and abnormal type of Brachiopod. In connexion
with this last view, it may be remarked that the operculum
of some of the species much resembles a valve of a
Brachiopod, and that the conical tube is in some of them
not a much greater exaggeration of the ventral valve of one
these shells than the peculiar *Calceola* of the Upper Silurian
and Devonian, which has been regarded by some palæontologists as a true Brachiopod. I have not, however, had any
opportunity of comparing the intimate structure of *Calceola*
with that of these shells. Shells of *Hyolithes* occur on the
Lower St. Lawrence in the beds with the phosphatic
nodules; and in one of these Mr. Weston has found a series
of conical shells of *Hyolithes* pressed one within another, as

---

* Canadian Naturalist, Dec. 1871.

† Mr. Matthew has directed my attention to the fact that Battande has
noticed that the shell of *Hyolithes* is similar in composition to the crust of
Trilobites.

if they had passed in an entire state through the intestine of the animal which produced the coprolite.

Inasmuch, then, as some of the most common invertebrates of the Cambrian seas secreted phosphatic shells, it is not more incredible that carnivorous animals feeding on them should produce phosphatic coprolites than that this should occur in the case of more modern animals feeding on fishes and other vertebrates.

We may now turn to the question as to the source of the abundant apatite of the Laurentian rocks. Were this diffused uniformly through the beds of this great system, or collected merely in fissure or segregation veins, it might be regarded as having no connexion with other than merely mineral causes of deposit. It appears, however, from the careful stratigraphical explorations of the Canadian Survey, in the districts of Burgess and Elmsley, which are especially rich in apatite, that the mineral occurs largely in beds of micaceous schist and limestone, interstratified with the other members of the series, though deposits of the nature of veins likewise occur. It also appears that the principal beds and veins are confined to certain horizons in the upper part of the Middle Laurentian, above the limestones containing *Eozoon*, though some less important deposits occur in lower positions.* This great apatite-bearing band of the Laurentian, consisting of beds of gneiss, limestone, and pyroxene-rock, has also been traced over a great extent of country east of the Ottawa river, and is well developed in the districts of Buckingham and Templeton. The mineral often forms compact veins or beds with little foreign intermixture; and these sometimes attain a thickness of several feet, though it has been observed that their thickness is variable in tracing them along their out-crops. The veins are especially rich in beds of pyroxenite, in which they form " fahlbands " or interrupted transverse veins thinning off on entering the gneiss or limestone. Thin layers of apatite also occur in the lines of bedding of the pyroxene-rock. In other cases, disseminated crystals are found throughout thick beds of limestone, sometimes, according to Dr. Hunt,

6

amounting to two or three per cent. of the whole mass.* Disseminated crystals also occur in some of the beds of magnetite, a mode of occurrence which, according to Dr. Hunt, has also been observed in Sweden and in New York, in the Laurentian magnetites of those regions. In one instance, at Ticonderoga, in New York, the apatite, instead of its usual crystalline condition, assumes the form of radiating and botryoidol masses, constituting tho Eupyrchroite of Emmons.

The veins of apatite fill narrow and usually irregular fissures; and the mineral is associated in these veins with calcite and with large crystals of mica. Since these veins are found principally in the same members of the series in which the beds occur, it is a fair inference that the former are a secondary formation, dependent on the original deposition of apatite in the latter, which must belong to the time when the gneisses, pyroxenites and limestones were laid down as sediments and organic accumulations.

In all the localities in which I have been able to examine the Laurentian apatite, it presents a perfectly crystalline texture, while the containing strata are highly metamorphosed; and this appears to be its general condition wherever it has been examined. Numerous slices of the more compact apatite of the beds have been prepared by Mr. Weston, of the Geological Survey; but, as might be expected, they show no trace of organic structure. All direct evidence for the organic origin of this substance is therefore still wanting. There are, however, certain considerations, based on its mode of occurrence, which may be considered to afford some indirect testimony.

If, with Hunt, we regard the iron ores of the Laurentian as organic in origin, the apatite which occurs in them may reasonably be supposed to be of the same character with the phosphatic matter which contaminates the fossiliferous iron ores of the Silurian and Devonian, and which is manifestly derived from the included organic remains.

* Vennor's Reports, 1872-73 & 1873-74.

* One of our Museum specimens is a nodule of apatite encased in a coating of graphite.

If we consider the evidence of *Eozoon* sufficient to establish the organic origin, in part at least, of the Laurentian limestones, we may suppose the disseminated crystals of apatite to represent coprolitic masses or the débris of phosphatic shells and crusts, the structure of which may have been obliterated by concretionary action and metamorphism.

Such Silurian beds of compact and concretionary apatite (without structure, yet manifestly of organic origin) as that described by Mr. Davies in the " Journal " of the Geological Society,* may be taken as representing the bedded apatite of the Laurentian. Further, the presence of graphite in association with the apatite in both cases may not be an accidental circumstance, but may depend in both on the association of carbonaceous organisms, whether vegetable or animal.

Again, the linguliferous sandstone of the Acadian group is a material which, by metamorphism, might readily afford a pyroxenite with layers or veins of apatite like those which occur in the Laurentian.

The probability of the animal origin of the Laurentian apatite is perhaps further strengthened by the prevalence of animals with phosphatic crusts and skeletons in the Primordial age, giving a presumption that in the still earlier Laurentian a similar preference for phosphatic matter may have existed, and, perhaps, may have extended to still lower forms of life, just as the appropriation in more modern times of phosphate of lime by the higher animals for their bones seems to have been accompanied by a diminution of its use in animals of lower grade.

The Laurentian apatite pretty constantly contains a small percentage of calcium fluoride; and this salt also occurs in bones, more especially in certain fossil bones. This may in both cases be a chemical accident; but it supplies an additional coincidence.

In the lowest portions of the Laurentian, no organic

* August, 1875.

remains have yet been detected; and these beds are
also poor in phosphates. The horizon of special prevalence of *Eozoon* is the Grenville band of limestone, which,
according to Sir William Logan's sections, is about 11,500
feet above the fundamental gneiss. It appears, from recent
observations of Mr. Vennor and Mr. W. T. Morris, that the
bed holding the Burgess *Eozoon* is on the same horizon
with the limestone of Grenville. The phosphates are most
abundant in the beds overlying this band. This gives a
further presumption that the collection and separation of
the apatite is due to some organic agency, and may indicate
that animals having phosphatic skeletons first became
abundant after the sea-bottom had been largely occupied
by *Eozoon.*

I would not attach too great value to the above considerations; but, taken together, and in connexion with the
occurrence of apatite in the Cambrian and Silurian, they
seem to afford at least a probability that the separation of
the Laurentian phosphate from the sea-water, and its accumulation in particular beds, may have been due to the
agency of marine life. Positive proof of this can be
obtained only by the recognition of organic form and
structure; and for this we can scarcely hope, unless we
should be so fortunate as to find some portion of the Lower
Laurentian series in a less altered condition than that in
which it occurs in the apatite districts of Canada. Should
such structures be found, however, it is not improbable
that they may belong to forms of life almost as much lower
than the *Lingulæ* and Trilobites of the Cambrian as these
are inferior to the fishes and reptiles of the Mesozoic.

Graphite of the Laurentian.*

In my paper of 1864, on the Organic Remains of the Laurentian
Limestones of Canada, as a sequel to the description of *Eozoon
Canadense*, I noticed, among other indications of organic matters in
these limestones, the presence of films and fibres of graphitic matter, and insisted on the probability that at least some of the lower
forms of plant life must have existed in the seas in which gigantic

* Quart. Journ. Geol. Soc. 1860.

Foraminifera could flourish. Dr. Hunt had previously, on chemical evidence, inferred the existence of Laurentian vegetation[†], and Dana had argued as to the probability of this on various grounds[‡]; and my object in referring to these indications in 1864, as well as to the supposed burrows of annelids, subsequently described by me[§], was to show that the occurrence of *Eozoon* was not to be regarded as altogether isolated and unsupported by probabilities of the existence of organic remains in the Laurentian deducible from other considerations.

Now that the questions which have been raised regarding *Eozoon* have been answered, not only by the adhesion of some of the greatest authorities in palæontology and zoology, but by the discovery of similar organisms in rocks of the same age elsewhere, and by specimens preserved in such a manner as to avoid all the objections raised to the mineral condition of the fossil, it may be proper to invite the attention of geologists more particularly to the evidence of vegetable life afforded by the deposits of graphite existing in the Laurentian.

The graphite of the Laurentian of Canada occurs both in beds and in veins, and in such a manner as to show that its origin and deposition are contemporaneous with those of the containing rock. Logan and Hunt state[†] that "the deposits of plumbago generally occur in the limestones or in their immediate vicinity, and granular varieties of the rock often contain large crystalline plates of plumbago. At other times this mineral is so finely disseminated as to give a bluish-gray colour to the limestone, and the distribution of bands thus coloured, seems to mark the stratification of the rock." He further states:—"The plumbago is not confined to the limestones; large crystalline scales of it are occasionally disseminated in pyroxene rock or pyrallolite, and sometimes in quartzite and in feldspathic rocks, or even in magnetic oxide of iron." In

---

† 'American Journal of Science' (2), xxxi. p. 305. From this article, written in 1861, after the announcement of the existence of laminated forms supposed to be organic in the Laurentian, by Sir W. E. Logan, but before their structure and affinities had been ascertained, I quote the following sentences:—"We see in the Laurentian series beds and veins of metallic sulphurets, precisely as in more recent formations; and the extensive beds of iron-ore, hundreds of feet thick, which abound in that ancient system, correspond not only to great volumes of strata deprived of that metal, but, as we may suppose, to organic matters which, but for the then great diffusion of iron-oxyd in conditions favourable for their oxydation, might have formed deposits of mineral carbon far more extensive than those beds of plumbago which we actually meet in the Laurentian strata. All these conditions lead us then to conclude the existence of an abundant vegetation during the Laurentian period."

‡ Manual of Geology. I may also be permitted to refer to my own work 'Archaia,' p. 168, and Appendix D, 1860.

§ Quart. Journ. Geol. Soc. vol. xxii. p. 608.

addition to these bedded forms, there are also true veins in which graphite occurs associated with calcite, quartz, orthoclase, or pyroxene, and either in disseminated scales, in detached masses, or in bands or layers "separated from each other, and from the wall rock by feldspar, pyroxene, and quartz." Dr. Hunt also mentions the occurrence of finely granular varieties, and of that peculiarly waved and corrugated variety simulating fossil wood, though really a mere form of laminated structure, which also occurs at Warrensburgh, New York, and at the Marinski mine in Siberia. Many of the veins are not true fissures, but rather constitute a net-work of shrinkage cracks or segregation veins traversing in countless numbers the containing rock, and most irregular in their dimensions, so that they often resemble strings of nodular masses. It has been supposed that the graphite of the veins was originally introduced as a liquid hydro-carbon. Dr. Hunt, however, regards it as possible that it may have been in a state of aqueous solution‡; but in whatever way introduced, the character of the veins indicates that in the case of the greater number of them the carbonaceous material must have been derived from the bedded rocks traversed by these veins, while there can be no doubt that the graphite found in the beds has been deposited along with the calcareous matter or muddy and sandy sediment of which these beds were originally composed.

The quantity of graphite in the Lower Laurentian series is enormous. In a recent visit to the township of Buckingham, on the Ottawa River, I examined a band of limestone believed to be a continuation of that described by Sir W. E. Logan as the Green Lake Limestone. It was estimated to amount, with some thin interstratified bands of gneiss, to a thickness of 600 feet or more, and was found to be filled with disseminated crystals of graphite and veins of the mineral to such an extent as to constitute in some places one-fourth of the whole; and making every allowance for the poorer portions, this band cannot contain in all a less vertical thickness of pure graphite than from 20 to 30 feet. In the adjoining township of Lochaber, Sir W. E. Logan notices a band from 25 to 30 feet thick, reticulated with graphite veins to such an extent as to be mined with profit for the mineral. At another place in the same district a bed of graphite from 10 to 12 feet thick, and yielding 20 per cent. of the pure material, is worked. When it is considered that graphite occurs in similar abundance at several other horizons, in beds of limestone which have been ascertained by Sir W. E. Logan to have an aggregate thickness of 3500 feet, it

† 'Geology of Canada,' 1868.
‡ 'Report of the Geological Survey of Canada,' 1866.

is scarcely an exaggeration to maintain that the quantity of carbon in the Laurentian is equal to that in similar areas of the
Carboniferous System. It is also to be observed that an immense
area in Canada appears to be occupied by these graphitic and
Eozoon-limestones, and that rich graphitic deposits exist in the
continuation of this system in the State of New York, while in
rocks believed to be of this age near St. John, New Brunswick,
there is a very thick bed of graphitic limestone, and associated
with it three regular beds of graphite, having an aggregate thickness of about 5 feet*.

It may fairly be assumed that in the present world and in those
geological periods with whose organic remains we are more familar
than with those of the Laurentian, there is no other source of un-
oxidized carbon in rocks than that furnished by organic matter, and
that this has obtained its carbon in all cases, in the first instance,
from the deoxidation of carbonic acid by living plants. No other
source of carbon can, I believe, be imagined in the Laurentian
period. We may, however, suppose either that the graphitic matter of the Laurentian has been accumulated in beds like those of
coal, or that it has consisted of diffused bituminous matter similar
to that in more modern bituminous shales and bituminous and oil-
bearing limestones. The beds of graphite near St. John, some of
those in the gneiss at Ticonderoga in New York, and at Lochaber
and Buckingham and elsewhere in Canada are so pure and regular
that one might fairly compare them with the graphitic coal of
Rhode Island. These instances, however, are exceptional, and the
greater part of the disseminated and vein graphite might rather be
compared in its mode of occurrence to the bituminous matter in
bituminous shales and limestones.

We may compare the disseminated graphite to that which we
find in those districts of Canada in which Silurian and Devonian
bituminous shales and limestones have been metamorphosed and
converted into graphitic rocks not dissimilar to those in the less
altered portions of the Laurentian.† In like manner it seems probable that the numerous reticulating veins of graphite may have
been formed by the segregation of bituminous matter into fissures
and planes of least resistance, in the manner in which such veins
occur in modern bituminous limestones and shales. Such bituminous veins occur in the Lower Carboniferous limestone and shale
of Dorchester and Hillsborough, New Brunswick, with an arrangement very similar to that of the veins of graphite; and in the

---

* Matthew in 'Quart. Journ. Geol. Soc.' vol. xxi. p. 423. Acadian Geolog,
p. 662.

† Granby, Melbourne, Owl's Head, &c., ' Geology of Canada,' 1863, p. 599.

Quebec rocks of Point Levi, veins attaining to a thickness of more than a foot, are filled with a coaly matter having a transverse columnar structure and regarded by Logan and Hunt as an altered bitumen. These palæozoic analogies would lead us to infer that the larger part of the Laurentian graphite falls under the second class of deposits above mentioned, and that, if of vegetable origin, the organic matter must have been thoroughly disintegrated and bituminized before it changed into graphite. This would also give a probability that the vegetation implied was aquatic, or at least that it was accumulated under water.

Dr. Hunt has, however, observed an indication of terrestrial vegetation, or at least of subaerial decay, in the great beds of Laurentian iron ore. These, if formed in the same manner as more modern deposits of this kind, would imply the reducing and solvent action of substances produced in the decay of plants. In this case such great ore beds as that of Hull, on the Ottawa, 70 feet thick, or that near Newborough, 200 feet thick,* must represent a corresponding quantity of vegetable matter which has wholly disappeared. It may be added that similar demands on vegetable matter as a deoxidizing agent are made by the beds and veins of metallic sulphides of the Laurentian, though some of the latter are no doubt of later date than the Laurentian rocks themselves.

It would be very desirable to confirm such conclusions as those above deduced by the evidence of actual microscopic structure. It is to be observed, however, that when, in more modern sediments, Algæ have been converted into bituminous matter, we cannot ordinarly obtain any structural evidence of the origin of such bitumen, and in the graphitic slates and limestones derived from the metamorphosis of such rocks no organic structure remains. It is true that, in bituminous shales and limestones, Macrospores and shreds of organic tissue can often be detected,† and in some cases, as in the Lower Silurian limestone of the La Cloche mountains in Canada, the pores of brachiopodous shells and the cells of corals have been penetrated by black bituminous matter, forming what may be regarded as natural injections, sometimes of much beauty. In correspondence with this, while in some Laurentian graphitic rocks, as, for instance, in the compact graphite of Clarendon, the carbon presents a curdled appearance due to segregation, and precisely similar to that of the bitumen in more modern bituminous rocks, I can detect in the graphitic limestones continu-

---

* Geology of Canada, 1863.
† See " Geological History of Plants," by the Author, page 48.

ous bands resembling elongated leaves or flattened stems*, fibrous structures which may be remains of vegetable tissue, and in some specimens vermicular lines, which I believe to be tubes of *Eozoon* penetrated by matter once bituminous, but now in the state of graphite.

When palæozoic land-plants have been converted into graphite, they sometimes perfectly retain their structure. Mineral charcoal, with structure, exists in the graphitic coal of Rhode Island. The fronds of ferns, with their minutest veins perfect, are preserved in the Devonian shales of St. John, in the state of graphite ; and in the same formation there are trunks of Conifers (*Dadoxylon Ouangondianum*) in which the material of the cell-walls has been converted into graphite, while their cavities have been filled with calcareous spar and quartz, the finest structures being preserved nearly as well as in comparatively unaltered specimens from the coal-formation†. No structures so perfect have as yet been detected in the Laurentian, though in the largest of the three graphitic beds at St. John there appear to be fibrous structures, which I believe may indicate the existence of land-plants. This graphite is composed of contorted and slickensided laminæ, much like those of some bituminous shales and coarse coals ; and in these there are occasional small pyritous masses which show hollow carbonaceous fibres, in some cases presenting obscure indications of lateral pores. I regard these indications however, as uncertain ; and it is not as yet fully ascertained that these beds at St. John are on the same geological horizon with the Middle Laurentian of Canada, though they certainly underlie the Primordial series of the Acadian group, and are separated from it by beds having the character of the Huronian.

There is thus no absolute impossibility that distinct organic tissues may be found in the Laurentian graphite, if formed from land-plants, more especially if any plants existed at that time having true woody or vascular tissues ; but it cannot with certainty be affirmed that such tissues have been found. It is possible, however, that in the Laurentian period the vegetation of the land may have consisted wholly of cellular plants, as, for example, mosses and lichens ; and if so, there would be comparatively little hope of the distinct preservations of their forms or tissues, or of our being able to distinguish the remains of land-plants from those of Algæ.

---

* These are abundant in some layers of limestone of the Grenville band at Lachute. Similar specimens have recently been described by Britton from the Laurentian of New Jersey, under the name *Archæophyton Newberrianum*,

† Acadian Geology, p. 535. In calcified specimens, the structures remain in the graphite after decalcification by an acid.

We may sum up these facts and considerations in the following statements:—First, that somewhat obscure traces of organic structure can be detected in the Laurentian graphite ; secondly, that the general arrangement and microscopic structure of the substance corresponds with that of the carbonaceous and bituminous matters in marine formations of more modern date; thirdly, that if the Laurentian graphite has been derived from vegetable matter, it has only undergone a metamorphosis similar in kind to that which organic matter in metamorphosed sediments of later age has experienced; fourthly, that the association of the graphitic matter with organic limestone, beds of iron ore, and metallic sulphides, greatly strengthens the probability of its vegetable origin; fifthly, that when we consider the immense thickness and extent of the Eozoonal and graphitic limestones and iron-ore deposits of the Laurentian, if we admit the organic origin of the limestone and graphite, we must be prepared to believe that the life of that early period, though it may have existed under low forms, was most copiously developed, and that it equalled, perhaps surpassed, in its results, in the way of geological accumulation, that of any subsequent period.

In conclusion, this subject opens up several interesting fields of chemical, physiological, and geological inquiry. One of these relates to the conclusions stated by Dr. Hunt as to the probable existence of a large amount of carbonic acid in the Laurentian atmosphere, and of much carbonate of lime in the seas of that period, and the possible relation of this to the abundance of certain low forms of plants and animals. Another is the comparison already instituted by Professor Huxley and Dr. Carpenter, between the conditions of the Laurentian and those of the deeper parts of the modern ocean. Another is the possible occurrence of other forms of animal life than *Eozoon* and Annelids, which I have stated in my paper of 1864, after extensive microscopic study of the Laurentian limestones, to be indicated by the occurrence of calcareous fragments differing in structure from *Eozoon*, but at present of unknown nature. Another is the effort to bridge over, by further discoveries similar to that of the *Eozoon bavaricum* of Gumbel, the gap now existing between the life of the Lower-Laurentian and that of the Primordial Silurian or Cambrian period. It is scarcely too much to say that these inquiries open up a new world of thought and investigation, and hold out the hope of bringing us into the presence of the origin of organic life on our planet. I would here take the opportunity of stating that, in proposing the name *Eozoon* for the first fossil of the Laurentian, and in suggesting for the period the name "Eozoic," I have by no means desired to exclude the possibility of forms of life which may have been precursors of what is now to us the dawn of organic existence. Should remains of still older organisms be found in those rocks now known to us only by pebbles in the Laurentian, these names will at least serve to mark an important stage in geological investigation.

IX. SUMMARY OF ARGUMENTS IN SUPPORT OF THE ANIMAL NATURE OF *EOZOON CANADENSE.*

1. It occurs in masses in limestone rocks, just as Stromatoporæ occur in the Palæozoic limestone.

2. While sometimes in confluent and shapeless sheets or masses, it is, when in small or limited individuals, found to assume a regular rounded, cylindrical or more frequently broadly turbinate form.

3. Microscopically it presents a regular lamination, the laminæ being confluent at intervals, so as to form a network in the transverse section. The laminæ have tuberculated surfaces or casts of such tuberculated surfaces, giving an acervuline appearance to those laminæ which are supposed to be the casts of chambers.

4. The original calcareous laminæ are traversed by systems of branching canals, now filled with various mineral substances, and in some places coarse and in many others becoming a fine tubulated wall. The typical form of these canals is cylindrical, but they are often flattened, especially in the larger stems.

5. In some specimens, large vertical tubes or oscula may be seen to penetrate the mass.

6. On the sides of such tubes, and on the external surface, the laminæ sub-divide and become confluent, thus forming a species of porous epidermal layer or theca.

7. Fragments of *Eozoon* are found forming layers in the limestone, showing that it was being broken up when the limestones were in process of deposition.

8. The great extent and regularity of the limestones show that they were of marine origin, and they contain

graphite, apatite and obscure organic (?) fragments other than Eozoon.

9. The ordinary specimens of *Eozoon* are mineralized with hydrous silicates (serpentine, &c.) in the same manner with Silurian and other specimens filled with glauconite, &c. These hydrous silicates also occur in the same limestones in concretions, bands, &c., in such a manner as to prove that they were deposited contemporaneously.

10. In some cases, the canals and chamberlets are filled with calcite and dolomite, in the manner of ordinary calcareous fossils, and this filling can often be distinguished from the original calcareous wall by a minutely granular or porous structure in the latter.

11. The specimens of *Eozoon* have been folded and faulted with the containing limestones, showing that they are not products of any subsequent segregation.

12. Similar testimony is borne by the fact that the masses of *Eozoon* are crossed by the veins of chrysotile which traverse the limestones and are of later origin.

13. The whole of the forms and structures seen in *Eozoon* correspond with those to be expected in a gigantic and highly generalised Rhizopod secreting a calcareous test, and possessing, as might be anticipated in such early organism, structures in some degree allied to such later forms as Stromatoporæ and calcareous sponges, which in the Eozoic it functionally represented.

14. The above evidence requires for its due appreciation the study of large suites of specimens in different states of preservation, and a practical knowledge of the different states of preservation of fossil remains, more particularly in the older and more crystalline rocks. Many objections taken have been based either on insufficient or imperfect specimens, or on want of the necessary experience in the study of the more ancient fossils in various states of preservation.

Finally, the question may be asked — What is the precise relation of *Eozoon Canadense*, considered as an animal organism, to any later and better known animals? This question may be answered in either of two ways :—(1) Functionally or in relation to the position of such an animal in nature; or (2) Zoologically, or with reference to its affinities to other animals. With reference to the first consideration, the answer is plain. The geological function of *Eozoon* was that of a collector of calcareous matter from the surrounding waters, then probably very rich in calcium carbonate, and its role was the same with that of the Stromatoporæ and calcareous Sponges, smaller Foraminifera and Corals in latter times. The answer to the second aspect of the question is less easy. An ordinary observer would at once place *Eozoon* with the Stromatoporidæ or Layer-corals, which fill or even constitute whole beds of limestone in the Cambro-Silurian, Silurian and Devonian Periods. While, however, Eozoon has been claimed on the highest authority for the Rhizopods, the Stromatoporæ and their allies have been regarded as Sponges, or more recently as Hydroids allied to the Hydractiniæ and Millepores.* I confess that I am not satisfied with these interpretations. I have in my collections large numbers of encrusting spinous forms, usually called Stromatoporæ, but which I have long set aside as probably Hydractiniæ. There are other forms with large vertical tubes which I have regarded as corals, but some Stromatoporæ seem to be different from either, and I am still disposed to regard many of them as Protozoa. Bearing in mind, however, that the Silurian is as remote from the Laurentian on the one hand as from the Tertiary on the other, we might be prepared to expect that if the Layer-corals of the Silurian are divisible into different groups, somewhat widely separated, we may be prepared to expect in the Laurentian much more generalised forms, less susceptible of classification in our modern systems. If, therefore, *Eozoon* were accessible to us in a living state, I should not be surprised to find that—while perhaps more akin to the calcareous-shelled Rhizopods than to any other modern group—it may have presented points of resemblance to Sponges or even to Hydroids, in its skeleton and

* See Nicholson's able memoirs, Publications of Pal. Socy., 1885.

mode of growth, and even in the structure of its soft parts. The precise facts as to these resemblances are, however, likely to be more or less uncertain; and in the meantime, the modern *Polytrema*, which encrusts shells and dead corals in the warmer seas, seems to me to present more resemblance to *Eozoon* than any other organism I know.

The following definition, originally in my little book, " Life's Dawn on Earth," may serve the purpose of characterising this remarkable form, which, whatever its nature, is certainly co-eval with and an agent in building up the limestones in which it occurs :—

EOZOON CANADENSE, Dawson.

General Form.—Broadly turbinate, often with a depression of cavity in the middle, sometimes rounded or depressed or confluent in sheets. Larger specimens often with several depressions or tubes penetrating the mass.

Structure of Test.—Calcareous, consisting of successive laminæ which are not continuous, but connect at intervals either by meeting together, by partitions, or by irregular pillars. Laminæ of granular texture, penetrated by minute tubuli, which in the thicker portions become branching canals. These are of rounded forms, sometimes articulated, and not unfrequently flattened at the extremities, or dividing into a vast number of tubuli.

Cavities of the Test.—Flattened, with slightly tuberculate surfaces, sometimes passing into series of rounded lobes, more or less circular (acervuline form) and occasionally sending off chamberlets into the thicker parts of the test. The canals terminate by a vast number of tubuli on the cavities, or sometimes by large tubes with widened openings. The chambers are larger and the walls of the test thicker and penetrated with larger canals near the base of the form. The upper layers, by separation of the constituent tubercles, often become acervuline.

The above structures are distinctly visible only in those specimens whose chambers and canals have been infiltrated with Serpentine, Pyroxene, Dolomite or Carbonaceous Limestone. Such specimens are sometimes associated with nodular masses of Serpentine or Pyroxene. Specimens in which the cavities have been filled with calcite, or obliterated by compression, rarely show any structure other than the fine granular texture or remains of the canals. Small and detached specimens, mineralised with Serpentine, are those most available for minute study.

X. CONCLUSION.

In concluding this resumé of facts and opinions respecting *Eozoon*, and of its representation in the Peter Redpath Museum, the author may be permitted to offer a few explanations, partly of a personal nature.

On my removal to Montreal in 1855, I had proposed to limit my geological work to two departments: (1) The completion of my researches in the Devonian and Carboniferous Geology and Palæobotany of the Maritime Provinces of British North America ; and (2) the study of the local phenomena and fossils of the Pleistocene deposits, as developed in the vicinity of my new home in the St. Lawrence Valley. My only concern with Laurentian geology on the one hand, and foraminiferal fossils on the other, lay in the obligation to know something of the former for teaching purposes, and to examine such forms of the latter as I found in the Pleistocene clays, and which led me to study the then recent publications of Carpenter, Williamson and Rupert Jones.

I took some interest in the discovery of Eozoon by Sir William Logan and his assistants, and it happened that I was the first to recognize its minute structures in some slices shown to me by Dr. Sterry Hunt, in connection with a paper which he was preparing on the mineralisation of fossil remains. I undertook the examination of the specimens at the request of Sir William Logan, and after offering to the late Mr. Billings, the Palæontologist of the Survey, to give him all the aid in my power if he would undertake the investigation. This, however, he declined, alleging the pressure of other work and his want of familiarity with microscopic research.

On the completion of my notes on the numerous specimens, not only of *Eozoon*, but of Laurentian and other crystalline limestones, submitted to me by Logan, I placed them with a number of camera drawings, prepared by the artist of the Survey, in the hands of Sir William, who was then about to proceed to England. Foreseeing the scepticism with which the announcement of Laurentian fossils was sure to be received, and not wishing to be involved in farther labour and controversy, I advised him to place my notes, along with the specimens and his own geological notes and those of Dr. Sterry Hunt on the mineralogical questions, in the hands of Dr. Carpenter and Prof. Rupert Jones, with *carte blanche* as to any use which these experts in the study of Foraminifera might be disposed to make of them. I had hoped that the matter was thus finally out of my hands, but the complicated and difficult questions which have since arisen, have made it a matter of obligation to devote more time to them than has been either agreeable or profitable. With the present publication I dismiss the matter finally, and without any feeling whatever as to the ultimate verdict of science with respect to these curious and puzzling specimens.

It is proper, however, to add that, independently of the nature of Eozoon itself, the questions it has raised have not been without advantage to Science. *Eozoon* is characteristic of Laurentian limestones in all parts of the world, and is not known in rocks later than those of the Archæan or Eozoic Period. For practical purposes, therefore, it is a Laurentian fossil. The matters discussed in the preceding pages show that it has directed attention to the nature and origin of the Laurentian beds, to the various modes of mineralisation of organic remains, to the structures of many ancient forms of animal life, as *Stromatopora, Hyolithes*, etc., to the origin and significance of the Laurentian Phosphates, to the origin of Graphite, and to the possible existence of

an Archæan flora. These and many other Geological questions have profited much by the discussion of Eozoon, and
for this at least we should be grateful.

It was hoped that my late lamented friend, Dr. W. B.
Carpenter, of London, who has done so much for the study
of *Eozoon*, and to whom in 1882 I had the pleasure of showing some of its best exposures, would have given to the
world a final and exhaustive memoir on the subject. I had
the pleasure of examining and discussing with him in 1884
many instructive and specially interesting slices which he
had prepared, and some of which had been photographed
under his direction. His latest letters show that he was
pursuing with unabated interest the study of his large collection of specimens. It is much to be desired that means
may be found for the publication of the materials in his
possession, if not for the completion of his work.

APPENDIX.

I. BIBLIOGRAPHY OF *EOZOON CANADENSE*.

The following is not intended as a complete Bibliography, but merely to afford means of reference to the original descriptions and the more important later contributions containing additional facts.

1. *Preliminary notices.*

After the original communication of Sir W. E. Logan to the American Association at Springfield in 1863, the notice by Sir W. E. Logan in the Geology of Canada, p. 48, 1863, and the preliminary statement by Sir W. E. Logan, Dr. Dawson and Dr. Hunt in the American Journal of Science of March, 1864, several short notices appeared in various quarters. Of these may be mentioned, Geological Magazine, July, 1864; British Association Report, in the same, vol. 1, p. 225 ; Bigsby's Laurentian Geology, Ib. p. 207; Letter by Dr. Carpenter to the President of the Royal Society, December, 1864; Geological Magazine, January, 1865.

2. *Original detailed descriptions, and notices subsequent thereto, in* 1865.

1. *February* 1*st*, 1865. On the occurrence of Organic Remains in the Laurentian Rocks of Canada, by Sir W. E. Logan, LL.D., F. R. S., &c., Quart. Journ. Geol. Soc. Vol. XXI, p. 45, with the following papers appended thereto.

(1) On the structure of certain Organic Remains in the Laurentian Limestones of Canada. By J. W. Dawson, LL.D., F. R. S., &c. Ibid. p. 51. 2 plates.

(2) Additional note on the structure and affinities of *Eozoon Canadense.* By W. B. Carpenter, M. D., F. R. S., &c. Ibid. p. 59. 2 plates.

(3) On the Mineralogy of certain Organic Remains from the Laurentian Rocks of Canada. By T. S. Hunt, Esq., M. A., F. R. S. Ibid. p. 67.

2. *April,* 1865. On the oldest known fossil, *Eozoon Canadense* of the Laurentian Rocks of Canada; its place, structure and significance. By T. Rupert Jones, F. G. S. Popular Science Review, Vol. IV., p. 343, &c., Pl. B.

3. *May,* 1865. On the structure, affinities and geological position of *Eozoon Canadense,* by W. B. Carpenter, M. D., F. R. S., &c. Intellectual Observer, No. X, May, 1865, p. 278. 2 plates.

4. *April,* 1865. On the History of *Eozoon Canadense* ; papers by Sir W. E. Logan, Dr. J. W. Dawson, Dr. W. B. Carpenter, and Dr. T. S. Hunt. (Same with No. 1, with additions). The Canadian Naturalist, new series, Vol. 7, p. 99, 1 plate.

5. *June,* 1865. Professors King and Rowney on *Eozoon,* stating objections to its organic nature. "The Reader," June 10th, 1865, p. 660.

3. *The more important papers, more especially by the original describers, arranged according to authors.*

1. *W. B. Carpenter on Eozoon Canadense.* Intellectual Observer, No. XL., p. 300, 1865.

———Supplemental notes on the structure and affinities of *Eozoon Canadense,* Quart. Journ. Geol. Soc. Lond. Vol. XXII, pp. 219–228, 1866.

——— —Notes on the structure and affinities of *Eozoon Canadense.* Canad. Nat., new ser., Vol. 11, pp, 111–119, wood cut, 1865. A reprint from Quart. Journ. Geol. Soc. Lond., 1865.

———Further observations on the structure and affinities of *Eozoon Canadense.* In a letter to the President. Proc. Roy. Soc. Lond. Vol. XXV., pp. 503–508, 1867.

———On the *Eozoon Canadense.* " Nature," vol. III, pp. 185, 186, 386, 1871.

———New observations on *Eozoon Canadense.* Ann., and Mag. Nat. Hist., ser. 4, vol. XIII., pp. 456–470, 1 plate, 1874.

———Final note on *Eozoon Canadense.* Ann., and Mag. Nat. Hist., ser. 4, vol. XIV., pp. 371–372, 1874.

———Remarks on Mr. H. J. Carter's letter to Prof. King on the structure of the so-called *Eozoon Canadense.* Ann. and Mag. Nat. Hist., ser. 4, vol. XIII, pp. 277–284 with 2 engravings, 1874.

———Remarks on *Eozoon Canadense.* " Nature," vol. IX, p. 491, 1874. (Abstract.)

———Further Researches on *Eozoon Canadense.* " Nature," vol. X, p. 390, 1874.

———On the replacement of organic matter by Siliceous Deposits in the process of Fossilization. " Nature," vol. X. p. 452, 1874. (Abstract.)

———Further Researches on *Eozoon Canadense.* Rep. Brit. Assoc. for 1874, pp. 136–137, 1875.

———New Laurentian Fossil. " Nature," vol. XIV., pp. 8, 9, 1876.

———Supposed new Laurentian Fossil. "Nature," vol. XIV., p. 68, 1876.

———Note on Otto Hahn's Microgeological Investigation of *Eozoon Canadense.* Ann. and Mag. Nat. Hist., ser. 4, vol. XVII., pp. 417–422, 1876.

———The *Eozoon Canadense.* " Nature," vol. XX, pp. 228–330, 1879.

———and J. W. Dawson. The *Eozoon Canadense.* " Nature," vol. XX., p. 328, 1879.

———*Eozoon Canadense.* The Microscope and its Revelations, sixth edition, pp. 587–592, 1881.

2. *J. W. Dawson* (and W. B Carpenter.) Notes on Fossils recently obtained from the Laurentian Rocks of Canada, and on objections to the organic nature of Eozoon. Quart. Journ. Geol. Soc. Lond.: vol. XXIII, pp. 257–265, 2 plates, 1865.

———Notes on Fossils recently obtained from the Laurentian Rocks of Canada, and objections to the organic nature of *Fozoon.* Amer. Journ. Sci., vol. XLIV., 2nd. ser., pp. 367–376, 1867.

———On certain organic remains in the Laurentian Limestone of Canada. Canad. Nat.. new ser., vol. XI., pp. 99, 111, 127, 128. 3 wood cuts, 1865

———Notes on Fossils recently obtained from the Laurentian Rocks of Canada, and on objections to the organic nature of *Eozoon,* with notes by W. B. Carpenter, M.D., F. R. S. Quart. Journ. Geol. Soc. Lond., vol. XXIII., pp. 257–265, plates XI, XII, 1867.

———(and W. B. Carpenter.) Notes on Fossils recently obtained from the Laurentian Rocks of Canada, and objections to the organic nature of *Eozoon.* Amer. Journ. Sci., vol. XLIV, 2nd. ser., pp. 367–376, 1867.

———(and W. B. Carpenter.) On new specimens of *Eozoon Canadense,* with a reply to the objections of Professors King and Rowney. Amer. Journ. Sci., vol. XLVI, 2nd ser., pp. 245–255, 2 plates, 1868

———Remarks on *Eozoon Canadense.* " Nature," vol. X., p. 103. 1 wood cut, 1874.

———Notes on the occurrence of *Eozoon Canadense,* at Cote St. Pierre. "Nature," vol. XII., p. 79, 1875. (Abstract.)

———On *Eozoon Canadense.* "Nature," vol. III., p. 287, 1871.

———The Story of the Earth and Man. London, 1873.

———The Dawn of Life: being the history of the oldest known fossil remains, and their relations to geological time and to the development of the animal kingdom, pp. 239, with 8 plates and 49 wood cuts. London, 1875.

——— On Mr. Carter's objections to *Eozoon.* Ann., and Mag. Nat. Hist., ser. 4, vol. XVII., pp. 118–119, 1876.

———Notes on the Phosphates of the Laurentian and Cambrian Rocks of Canada. Quart. Journ. Geol. Soc. Lond., vol. XXXII., pp. 285 -291, 1876.

——— Notes on the occurrence of *Eozoon Canadense* at Cote St. Pierre. Quart. Journ. Geol. Soc. Lond., vol. XXXII, pp. 66–74, plate X. with 4 wood cuts, 1876.

——On some new specimens of Fossil Protozoa from Canada. Proc. Am. Assoc. Adv. Sci. XXIV., pp. 100–106, wood cuts, 1876.

——New Facts relating to *Eozoon Canadense.* Proc. Am. Assoc. Adv. Sci., vol. XXV., pp. 231–234, 1876.

——*Eozoon Canadense*, according to Hahn. Ann. Mag. Nat. Hist., ser. 4, vol. XVIII., pp 29–38, 1877.

——New Facts relating to *Eozoon Canadense.* Canad. Nat. new ser., vol. VIII., pp. 282–285, 1878.

——- On the Microscopic Structure of Stromatoporidae, and on Palæozoic Fossils mineralized with Silicates, in illustration of *Eozoon.* Quart. Jour. Geol. Soc. Lond., vol. XXXV., pp. 48–66, 3 plates, 1879.

——Möbius on *Eozoon Canadense.* Amer. Journ. Sci., vol. XVII., p. 196, wood cuts, 1879.

——Note on recent controversies respecting *Eozoon Canadense.* Can. Nat., vol. IX., p. 228, 1879.

——Notes on *Eozoon Canadense.* The Cana. Rec. of Sci., vol. I., pp. 58–59, 1884.

——On the Geological Relations and Mode of Preservation of *Eozoon Canadense.* Report Brit. Assoc, (Southport, 1883.) p. 494, 1884.

——Canadian and Scottish Geology, an address delivered before the Edinburgh Geological Society at the close of session, 1884. Trans. Edin. Geol. Soc., vol. V., pp. 113–114, 1885.

——New Facts Relating to *Eozoon Canadenses.* Geol. Maga., February, 1888.

3.   *C. W. Gümbel.* Ueber das Vorkommen von *Eozoon* in dem ostbayerischen Urgebirge; Sitzungsber. d. k. v. Akad. Wiss. Munch., 1866, Bd. I, pp. 25–144, 3 plates.

4.   *T. S. Hunt.* Laurentian Rhizopods of Canada. Extract of a letter from T. Sterry Hunt, F. R. S., to J. D. Dana, April 2nd, 1864. Amer. Journ. Sci.,vol. XXXVII, 2nd. ser. p. 431, 1864.

——On the Mineralogy of *Eozoon Canadense.* Canad. Nat. n. s., vol. II., pp. 120–127, 1 plate, 1865.

—————On the Mineralogy of certain Organic Remains from the Laurentian Rocks of Canada. Quart. Journ. Geol. Socy. Lond., vol. XXI., pp. 67-71, 1865.

—————Geology and Mineralogy of the Laurentian Limestones. Geological Survey of Canada. Report of progress from 1853 to 1866, pp. 181--233. Ottawa, 1866.

—————The Geological History of Serpentines, including Notes on pre-Cambrian Rocks. Trans. Roy. Soc., Canada, vol. I., pp. 165--215, 1883.

T. R. Jones. *Eozoon Canadense* in this country. Nat. Hist. Rev., Lond., vol. V., pp. 297–298, 1865.

—————On the oldest known fossils, *Eozoon Canadense* of the Laurentian Rocks of Canada; its place, structure and significance. Popular Sci. Review, 1867, pp. 343–352, plates XV. and two wood cuts.

W. E. Logan. Supposed Fossils in the Laurentian Limestone. Geology of Canada, pp. 48--49, 2 wood cuts, 1863.

—————On Organic Remains in the Laurentian Rocks of Canada; Amer. Journ. Sci., vol. XXXVII., 2nd series, pp. 272–273, 1864.

—————On organic remains in the Laurentian, Journ. Geol. Socy., 1865.

—————(J. W. Dawson and T. S. Hunt.) On the occurrence of Organic Remains in the Laurentian Rocks of Canada. Report Brit. Assoc. (Bath meeting), Trans. Sections, p 225, 1864.

—————On new Specimens of *Eozoon.* Quart. Journ. Geol. Soc. London, vol. XXIII., pp. 253–257, 1867.

II. SYNOPSIS OF SPECIMENS OF *EOZOON* IN THE PETER REDPATH MUSEUM.

1. *Specimens showing general form.*

Three specimens from Côte St. Pierre, weathered out, showing broad unequally turbinate forms.

Several specimens showing other forms weathered, and showing rounded and flattened shapes.

Large confluent specimens indicating aggregation of simple forms.

Specimen with two tubes or oscula penetrating it, and showing modifications of laminæ on approaching the tubes.

Other specimens showing what seem to be wider oscula.

Large convex specimen based on Pyroxene –the specimen figured in Life's Dawn on Earth.

Specimens contorted by the movements of the containing beds.

Photograph of elongated or clavate specimen from the Hastings group.   Original in Ottawa Museum.

Thin and irregular specimens with only a few layers.

Specimens showing forms apparently calcified rather than serpentinous, and showing less distinct layers than those with serpentine.

2. *Specimens showing structure.*

Laminated specimens, weathered, etched and polished, showing many varieties of the structure. Some of these are Burgess specimens, mineralized with Loganite.

Specimens polished and etched and showing the laminæ, and in some cases the canals. These are merely mounted in glass topped boxes. Among them are specimens showing pyroxene and dolomite, filling canals and chambers, and specimens traversed by chrysotile veins of secondary origin.

Variety showing continuous laminæ with intervening canals or vesicles in manner of Cryptozoum of Hall.

Variety *minor*, with very narrow chambers and thin laminæ.

Acervuline variety and acervuline parts of large masses.

3. *Specimens showing states of preservation and modes of occurrence.*

(1.) The greater part of the specimens are mineralized with serpentine, but this is of different qualities, more especially deep green, light green and amber-coloured and cream-coloured varieties, and the ferruginous variety when fresh, nearly colourless but weathering reddish. Most of the above are from Petite Nation and Grenville·

(2.) Other specimens from Burgess, mineralized with Loganite.

(3.) Large slabs in upright cases from Logan collection show the alternation of beds in the *Eozoon* limestones, also layers of calcite which when etched are seen to be full of fragments of Eozoon. A fragment etched (in the horizontal case) shows several layers, some of them containing fragmental *Eozoon.*

(4.) Fragments of *Eozoon* in separate chamberlets—formerly called Archæosphærinac. There are several specimens showing these.

(5.) *Eozoon* mostly fragmental, from Chelmsford, Massachusetts, from the Laurentian of New York, from the Alps and from Brazil.

(6.) Large polished specimen from Logan collection, showing contortion and the more massive form.

(7.) Large weathered and other specimens from Petite Nation showing irregular forms. One of these is based on pyroxene, and is of the kind which has been described as a mere rock of transition.

4. *Summary of states of preservation illustrated by the specimens.*

(1.) Mineralized with light or nearly white serpentine.

(2.) Mineralized with ferruginous serpentine, weathering rusty.

(3.) Mineralized with dark green serpentine.

(4.) Mineralized with pyroxene, or partly with pyroxene and partly with serpentine.

(5.) Mineralized with Loganite.

(6.) Specimens traversed with chrysotile veins.

(7.) Fragmental, in large slabs of Logan collection and in smaller specimens etched or in natural state.

(8.) Weathered specimens, showing the detached cellules known as as Archæospherinæ.

(9.) Weathered specimens showing various appearances in different stages of erosion.

5. *Associates of Eozoon.*

(1.) Cylindrical casts in pyroxene, with central cores of pyroxene and traces of radiating laminæ, Laurentian, Ottawa District.

(2.) Bedded graphite and graphitic limestone and graphitic gneiss, graphite from veins.

(3.) Limestone with apatite, and apatite associated with graphite, Petite Nation.

(4.) Limestones, gneiss, &c., associated with the Grenville band of limestone.

6. *Imitative forms.*

   (1.) Banded trap, pyroxene and felspar, Montreal Mountain.
   (2.) Banded gneiss, Laurentian.
   (3.) Banded limestone, Laurentian.
   (4.) Banded apatite, Laurentian.
   (5.) Banded quartz and tourmaline, Laurentian.
   (6.) Layers of calcite and serpentine, Laurentian.
   (7.) Graphic granite and other similar rocks.
   (8.) Serpentine grains, disseminated in limestone.
   (9.) Chrysotile veins in serpentine.

The above are in trays below the table-case with *Eozoon.*

7. *Microscopic specimens, &c.*

   A cabinet containing extensive series of slices, transparent
or etched, of *Eozoon* and of various fossils similarly mineralized
with silicates, as well as specimens from which the slices have
been taken, is placed in the Board-room of the Museum and
will be accessible to those desiring to study such specimens.
This collection contains a variety of examples of different
states of preservation and kinds of mineralization, and includes
the specimens originally prepared and studied by Sir Wm.
Dawson.

CONTENTS.